LIV

ROSSIER, Day

Living with chickens

LIVING
with
CHICKENS

Everything you need to know about keeping chickens

jay rossier

D&C
David and Charles

A DAVID & CHARLES BOOK
David & Charles is a subsidiary of F+W (UK) Ltd.,
an F+W Publications Inc. company

This edition published in the UK by David & Charles Ltd in 2005
First published in the USA by The Lyons Press in 2002

Text copyright © Jay Rossier 2002, 2003, 2005
Photographs copyright © Geoff Hansen 2002, 2003, 2005

Jay Rossier has asserted his right to be identified as author of this
work in accordance with the Copyright, Designs and Patents Act, 1988.

A catalogue record for this book is available from the British Library.

ISBN 0 7153 2198 6 paperback

Printed in Singapore by KHL Printing
for David & Charles
Brunel House   Newton Abbot   Devon

Visit our website at www.davidandcharles.co.uk

David & Charles books are available from all good bookshops; alternatively you can
contact our Orderline on (0)1626 334555 or write to us at FREEPOST EX2 110,
David & Charles Direct, Newton Abbot, TQ12 4ZZ (no stamp required UK mainland).

To Andy, Sam, and Liam

CONTENTS

Opposite: An apple tree frames Tom Powers' chicken house in South Royalton, Vermont. Built in 1983, the building is too far from the main house, according to Powers: "I've had to dig a 100-foot path through the snow ten times some winters."

ACKNOWLEDGMENTS

FOR MANY REASONS THIS BOOK OWES ITS EXISTENCE to the patience, persistence, and good humor of Ann Treistman of The Lyons Press. Many thanks also to Geoff Hansen for his photographs and friendship, and The Mountain School of Milton Academy in Vershire, Vermont, for the use of their space.

Many friends and acquaintances have shared their chicken experiences with me and helped in other ways, but I had particularly valuable assistance from Gerry Coleman, Alex Keats, Bob Machin, and Mary Hays.

Finally, many thanks to MaryAnn Dubé for her design, and to Kevin Lynch and Chris Mongillo at Lyons for their hard work on the production of this book.

—JAY ROSSIER

THANKS ARE DUE TO CHICKEN FARMERS Carrie Maynard, Tom Powers and Lianne Thomashow, the Vermont Bird Fanciers Club, author Jay Rossier, and especially editor Ann Treistman at The Lyons Press.

—GEOFF HANSEN, PHOTOGRAPHER

Opposite: One of Lianne Thomashow's birds, spending an afternoon in the chicken coop on a cold afternoon.

INTRODUCTION

Welcome to the wonderful world of poultry in general, and chickens in particular. Jay Rossier is right on target when he says, "Chickens have a charm that will affect even those with no bird experience." That was me—as a young husband and father in the 1970s, I had absolutely no bird experience.

How I wish we had had a book like this when we first started with poultry. Our involvement with chickens as a family began with our four-year-old daughter's prayer, "Thank you, God, for the milk we get at the store, and the eggs we get at the store…" Determined to help Shara and her little brother learn where eggs came from, I purchased six day-old chicks from a flock owner who had placed an ad in our local paper. Wouldn't you know it, four of the six turned out to be roosters, but Mandy and Mindy, two Ameraucana pullets, became important members of our first flock. At about the same time, a friend gave us a dozen fertile eggs, which we placed into a homemade incubator—a small aquarium covered with a piece of plywood that had a light bulb and a thermostat mounted on it. Even as an absolute novice, with that contraption I got three of twelve eggs to hatch. "Daddy, one of the chicks just has one eye!" our daughter exclaimed on seeing the results of our labors. She promptly named it "Charlie One-Eye" after

Opposite: Columbiana, a Columbian Wyandotte hen, takes a stroll in the fresh snow in the pen the chickens share with Carrie Maynard's horses in South Royalton, Vermont.

the Charlie who provided the eggs. Much to our daughter's chagrin, Charlie One-Eye turned out to be a girl, and an excellent layer. Next, I ordered twenty-five day-old pullets from the Murray McMurray hatchery. We were off and running in the chicken business.

At that time we lived in upscale suburbia, but I could and did meet all city health codes. I was determined to be a good neighbor, so we kept no roosters, and I made sure that there were no rodents or nasty odors to offend our neighbors on three sides. Fascinated by all the different breeds we saw in the hatchery catalogues, we kept saying, "Oh, we just *must* have some of those." By now, our flock was providing far more eggs than our family could consume in a healthy way. What to do?

At first we gave away our excess eggs to friends, who clamored for more. I quickly saw the potential for our daughter's first exposure to capitalism. Her mother purchased a small egg scale (which now proudly decorates a shelf at the farm), and Shara set about to weigh each egg, selling them by weight, just like the grocery store. Her trademark was one colored egg from Mandy or Mindy (Ameraucanas usually lay green or sometimes blue) in each dozen. Some folks bought a dozen just to get that colored egg to show their neighbors and friends. We always had more buyers than eggs to sell. And I shall never forget the day our daughter said to me thoughtfully, "Daddy, we need to take in a little more money than we have to pay out for feed." Our six-year-old had learned through life experience what I had paid big bucks to learn in Economics 101 at college. Our flock grew to as many as fifty-five pullets and hens in suburbia.

On a whim we entered one of those McMurray pullets, a Bearded Buff Laced Polish, in a local show. She won Best of Breed and Champion Continental Class. As I write these words I am looking fondly at the trophy that "Goldie" won at the show. That fall we entered her in our Texas State Fair

Poultry Show, and she won again. Now we were hooked, not only on backyard poultry for meat and eggs, but on standard-bred poultry for exhibition as well.

At these two poultry shows we met many nice folks, dedicated poultry breeders, who introduced us to the

Above: This breed is officially called New Hampshire, but in practice people often call it New Hampshire Red because it was developed from the Rhode Island Red—and it *is* red. According to the New Hampshire Breeders Club of America, in 1935 the club applied to have the breed admitted into the American Standard of Perfection under the name of New Hampshire Red. The Rhode Island breeders objected to including the word "Red," so it was admitted under the name "New Hampshire" instead.

American Poultry Association and a book called *Standard of Perfection*. The standard describes each bird of every breed and color variety in minute detail, and poultry judges use that standard, which details the perfect bird, when judging birds in competition. We also learned that, in addition to strong poultry clubs in both Fort Worth and Dallas, there were poultry clubs

Above: Ebenezer, who is a rare breed Suffolk Punch horse, shares his hay with a Dark Brahma hen in Carrie Maynard's barnyard. Maynard said she's had far fewer flies and mosquitoes bothering her horses since getting chickens three years ago.

in Abilene, Waco, Cleburne, Terrell, and Wichita Falls that sponsored one or two poultry shows each year, and all of these were within easy driving distance of our home.

By that time, we had moved south of town to an acre in the country. The first order of business had been to build a poultry building thirty-six feet long by twelve feet wide and divide it with chicken wire into nine pens, each four feet wide. Over a few months I had selected Bearded White Silkies, White Crested Black Polish, and Single Comb White Leghorns as my breeds of choice, and purchased two or three breeding trios (a trio is two females and a male) of each variety. The homemade incubator gave way to a one-hundred-egg-still-air, which gave way to a three-hundred-egg-forced-air, automatic-turning incubator.

Above: A Columbian Wyandotte hen goes on the lookout for seeds, insects, and pebbles in the woods.

What had started out as a plan for six to ten layers to provide fresh eggs turned into breeding and exhibiting chickens at ten to fifteen shows a year, and hatching four to five hundred bantam chickens each year.

The first poultry show for the Malones was more than thirty years ago, and I am still hooked on chickens, now completing my second two-year term as

President of the American Poultry Association, Inc., the oldest continuous livestock organization in North America, founded in 1873 at Buffalo, New York.

Our family has great memories of our time together with the birds, whether breeding and raising chickens at home, or showing the grown birds in competition. Our son and daughter eventually went away to college and have their own lives now. My wife says that two of her children outgrew the chickens, and the third one (guess who) is still fooling around with them. The grandkids head straight for the incubator when they arrive at Grandma and Pawpaw's house to see what is inside. You can't tell them it is empty. They have to see for themselves.

If you are interested in learning more about the American Poultry Association, Inc. and the resources we have that might be of interest and help to you, I invite you to visit our website at www.ampltya.com. My involvement with the APA has led me from backyard flock owner to breeder and exhibitor of standard bred chickens, to general licensed poultry judge, to being elected the 40TH president of the APA in March, 1998. If you are looking for an animal project for your children or grandchildren but have limited space and financial resources, a trio of standard bred chickens or a meat pen of broilers are excellent options. The kids will learn

responsibility and accounta-
bility just as well from chickens as they would from
caring for a calf, pig, or
sheep. The APA/ABA Joint
Youth Program is outstand-
ing, in my biased opinion.
Please let us know if the APA
can be of service to you in
any way.

It is a privilege to be
associated with Mr. Rossier
and Lyons Press in bringing
this volume to novice poultrymen and -women. I pre-
dict your chickens will bring
you as much meaning and
joy as Mindy and Mandy and
Charlie One-Eye and all the
others have to two generations of our family. I hope
you raise some excellent
birds whether for meat, eggs,
exhibition, or all of the
above. Best wishes and much success with your flock!

PAT MALONE, PRESIDENT
AMERICAN POULTRY ASSOCIATION, INC.

Above: This hen has a new owner at a sale and swap hosted by the Vermont Bird Fanciers Club at the East Randolph Community Hall in East Randolph, Vermont.

CHAPTER ONE
The Charm of Chickens

"You are not keeping them, of course, to make or even to save money. You are not keeping them as pets.

You are keeping them for the simple pleasure of their company and the beauty and tastiness of their eggs and their meat. You are raising them because you wish to strike a modest blow for the liberation of the chicken—and, indeed, of all living things on earth."

—CHARLES DANIEL AND PAGE SMITH,
The Chicken Book, 1975

It wasn't my idea, in the beginning, to start raising chickens. I was living thirty miles from a nearby university town in an old farmhouse with a lot of land and a couple of unused outbuildings. A medical student friend of mine who lived in cramped student housing not only needed an excuse to get out, but thought that a livestock project might make him feel more in touch with what to him seemed like the real world: the country. He provided the inspiration and the mail-order catalog from one of the hatcheries; I had the space, some livestock experience, and a willingness to bring food and water to the mysterious creatures on a daily basis. Chickens

Opposite: A White Leghorn hen struts across the yard at Tom Powers' home.

Above: A bantam Old English Game hen struts through the woods. For nearly every large breed chicken, there is a bantam equivalent.

seemed to me then to be stupid, fearful, and aggressive. They are full of sharp points from their beaks to their toes and move in a distinctly jagged way, jerking their heads more like a reptile than a bird. In the farm-animal department, I liked cows, which are massive and deliberate. And warm.

But chickens have a charm that will affect even those with no bird experience. In short order I began to appreciate the rich colors and textures of their various plumages, their weight and shape. They are stately, dignified, and industrious creatures that take their work of scratching and eating and laying and setting seriously. Furthermore, they have a genuine, if somewhat detached, curiosity about us, and are happy to work alongside us in whatever we busy ourselves with outside.

Of course, the eggs and meat they provide is superior to what you can get from the store. If what you want is home-grown animal protein, you'll soon discover that these birds can offer it—and that they are a lot cheaper and easier to house, feed, herd, and transport than sheep, goats, pigs, cows, ostriches, or what have you.

Before you get started, make sure that there are no local zoning laws that might end your career in chicken husbandry before it even begins. Make a call to your town clerk or city council to find out the regulations in your town. In addition, it's always a good idea to broach the subject with close neighbors before diving in.

WHAT KINDS OF CHICKENS SHOULD YOU KEEP?

Your choice of what kinds of birds to keep depends first and foremost on whether you want meat or eggs, or if you are simply buying them for yard ornamentation. Some are more appropriate for meat, some are better for eggs, and some were bred to do both tolerably well. Once you know what you want from your chicken, you can begin to imagine some of its characteristics: size, temperament, and looks. There are chickens bred to be attractive for showing (some of those get to be pretty silly looking, although this is, of course, a matter of taste). There are breeds suited for cold weather, others which prefer warm; some with relaxed dispositions, and others that can be nasty but delicious. You should take these factors into account when deciding which chicken is right for you. When choosing your breed, don't be surprised to hear fellow poultrymen talk about "the Standard." They are referring to the American Poultry Association's publication, *Standard of Perfection*, which describes each breed in detail. The Standard is used in judging at poultry shows and to help chicken breeders improve their flocks over time by breeding for preferred characteristics.

A breed is a group of related chickens that has the same general size and shape; shares the same skin color, number of toes, and plumage style; and has the same style of comb, which is the fleshy, spiky, red topknot on the chicken's head.

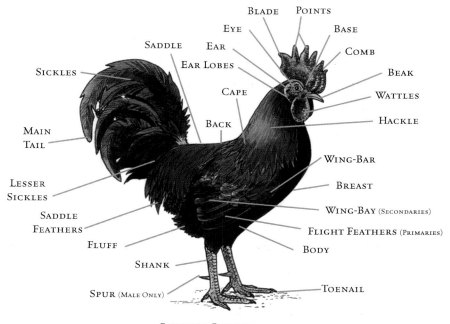

BLADE   POINTS
EYE   BASE
SADDLE   EAR   COMB
EAR LOBES
SICKLES   BEAK
CAPE   WATTLES
HACKLE
BACK
MAIN
TAIL
WING-BAR
LESSER
SICKLES   BREAST
SADDLE   WING-BAY (SECONDARIES)
FEATHERS   FLIGHT FEATHERS (PRIMARIES)
FLUFF   BODY
SHANK
SPUR (MALE ONLY)   TOENAIL

PARTS OF A CHICKEN

In order to talk about different breeds of chicken, it might be useful to know the names of the parts of the chicken. Above is a diagram showing the parts of a chicken used when people are talking about what makes one breed different from another.

MEAT, EGGS, OR BOTH?

Some breeds have been selectively bred over the years to emphasize meat-producing characteristics. Others have been selectively bred to be exceptionally prolific egg layers. The females of the meat breeds will certainly lay eggs, but perhaps as many as fifty fewer in a year than a chicken of an egg-laying breed. Laying chickens can be butchered for meat, but they may consume twice as much feed over a longer period of time before gaining the same weight as a corresponding meat bird, and the result will not be as tender as a meat bird's meat.

It is wise therefore to raise laying breeds for eggs and meat birds for meat. A third option, however, exists in dual-purpose breeds that were popular 100 years ago when the economics of small-scale agriculture required breeds in which the hens would lay many eggs but their male offspring would flesh out as quickly and efficiently as possible.

MEAT BIRDS

Jersey Giant, Brahma, Cochin, and Cornish are a few examples of meat breeds. The bird bred for modern factory farming, however, is called the Cornish-Rock Cross or Rock-Cornish Cross, which is the product of a Rock mother and a Cornish father. The Cornish-Rock Crosses are large birds with huge appetites and little interest in getting any exercise. A chicken that walks around scratching and exploring very much is using feed energy for those activities instead of for making meat. These birds grow to a good size (about 4 pounds) in a short time (from 6 to 8 weeks) and are very efficient converters of feed to meat: about 2 pounds of feed for every pound of meat. They are therefore the cheapest means of producing frying or roasting chickens for your freezer. Also, because their feathers are exclusively and entirely white, their skin is cleaner looking than colored birds that may appear to have black dots all over them after they are plucked.

LAYING HENS

Hens bred especially for laying are designed to put their energy into eggs and not into body mass, and so are somewhat smaller than birds bred for meat. They begin laying a little sooner than the meat or dual-purpose breeds—at about 5 months of age

Opposite: Two pullets— a Silver-Laced Wyandotte, at left, and Golden-Laced Wyandotte— take a look at the fresh snow and decide to stay in the coop at Carrie Maynard's barnyard

instead of 6. White Leghorn is the most common and most productive modern laying hen. A Leghorn will consume less grain per dozen eggs and will produce more eggs over the course of a year than the dual-purpose or meat breeds. When they are about 18 months old, chickens begin to shed their feathers, or molt, in order for new ones to grow in. During this time they will stop laying for a few weeks to a few months. Modern egg-laying breeds have been bred for shorter molting periods and thus produce more eggs in a year.

DUAL-PURPOSE BREEDS

The dual-purpose breeds are what we now think of as the old-time breeds—such as New Hampshire and Rhode Island Red—that were developed in England and America in the 18th and 19th century. At that time (and right on into the middle of this century), 80 to 90 percent of American households kept chickens to supplement their diet and their annual cash income, as it was easy to sell extra eggs locally. Before mass-production methods came to agriculture the goal was a hen that laid well for as long as possible and that produced offspring that would flesh out well for the stew pot or roasting oven.

The hens of the dual-purpose breeds—such as New Hampshire, Rhode Island Red, Orpington, Wyandotte, Dominique, and Plymouth Rock—do not lay as many eggs in a year as the Leghorn or other single-purpose egg-laying breeds. The cockerels, or male chicks, don't gain weight as quickly or as cheaply as the Cornish-Rock Cross or other single-purpose meat breeds. But the dual-purpose breeds are attractive to many backyard poultry enthusiasts because they are sturdy and attractive, and are more self-sufficient than the single-purpose breeds—meaning that they are more willing to find food for themselves if they have room to roam outdoors. They also have the nostalgic appeal of a time before modern industrial agriculture. Because they aren't productive enough to be used in modern mechanized poultry production, many are considered endangered species and need to be used productively in order to be saved from extinction.

SIZE

This is the greatest distinction that separates one breed of chicken from another. A breed is either a "bantam" breed or it is not a "bantam" breed, in which case it is a "large" breed. The large breeds, simply because they are large and therefore produce greater quantities of meat and eggs, have historically been given more attention in terms of breeding for production. Presumably because they are miniature and therefore cute, bantams are thought of as show birds first, although pound for pound they can produce as well as many of the large breeds.

Occasionally large breeds are referred to as "standard" breeds. This can lead to confusion on the part of beginners who will more often hear the phrase "the Standard" used as shorthand for the American Poultry Association's Standard of Perfection, or the description of a particular breed in it, or another standard published by a particular breed organization.

Large Breeds

Large breeds produce more meat and eggs than bantams both because they are bigger and because they have been bred to produce more meat and more and larger eggs. The reasons to keep a large breed over a bantam are entirely a matter of taste. People who keep large chickens instead of bantams are the types that would choose a retriever or German shepherd over a springer spaniel.

Bantams

There have probably always been smaller chickens, and a bird resembling the Silkie breed of bantam chicken turns up in the log of Marco Polo in the 13th century. The name bantam is from Bantam Island in the Dutch East Indies, which was a meeting point for the trade routes between Asia and the west in the 17th century. The native fowl of the region were used by the sailors as a source of meat and eggs during their voyages, perhaps because smaller birds were easier to keep in the smaller living spaces aboard ship. It was common then to call any small birds "bantams," whether they originated on Bantam Island or not.

Bantams are anywhere from one-quarter to one-fifth the size of a large breed. Although there are a number of "true" bantam breeds without full-sized counterparts, in many cases bantams are miniaturized versions of large breeds developed during the 19th and 20th centuries in America and Europe. Bantams eat less and are both willing and able to forage for grass, bugs, worms, garbage, and anything else in their line of sight. The eggs of bantams are quite noticeably smaller, but there is nothing wrong with small eggs if you are eating them yourself and don't need to sell them through a distributor in order to make a living (as your counterpart in the 1940s and 1950s did). About three bantam eggs will make as big an omelet as two regular-sized eggs. On the flip side, bantams fly throughout their lifetime, unlike the larger breeds, which lose much of their ability and interest in flying when they mature. It is therefore harder to get bantams to go just where you want them to go. They are also a little more high-strung than large breeds.

LOOKS

Part of the charm of chickens is certainly the vivid coloring and patterning of their plumage and their various and occasionally spectacular configurations of comb and plumage style. The white-feathered birds are cleaner looking when dressed for the oven than colored-feathered birds, and are somewhat less interesting to look at while they are alive—but they are

Opposite: A Silver-Laced Wyandotte pullet roams the yard. A pullet is a young female chicken who is not yet a year old.

Above: A Golden-Laced Wyandotte pullet, left, and Silver-Laced Wyandotte hen pull up short as they come face-to-face while roaming the woods behind their coop.

not around very long anyway. You are going to be living with your laying hens every day for a period measured in years, so choose a breed that is attractive to you. The colors are rich and pleasing on any healthy bird, whether a solid-colored Rhode Island Red or a richly patterned Silver-Penciled Wyandotte. Those with a taste for the more gaudy have plenty of options, too. Frizzles, named for their scrub-brush–like appearance, have feathers that curl back toward the bird's head instead of lying flat and pointing toward the tail. Silkies have fluffy downlike

feathers that make the birds look like animated powder puffs or feather dusters, or as if they have hair like cats, as Marco Polo wrote. Some chickens have various styles of feather headdresses and top-knots; others have feathered boots covering their feet. When it comes to looks, there is a breed of chicken for everyone.

HARDINESS AND TEMPERAMENT

In cold climates, the heavier and more heavily feathered breeds—such as the Rocks, Reds, Orpingtons, and Brahmas—may produce a little better and further into the winter than smaller breeds. Raise your meat birds in the summer so you don't have to worry about the cold. Construct your coop carefully to keep out drafts. In these ways you can make cold hardiness a nonissue and raise any sort of bird you please successfully.

Hardiness, on the other hand, is also a function of attitude on the part of a chicken, and the birds bred in recent decades specifically for industrial egg and particularly meat production have had some of the Protestant virtues of independence, thrift, and self-sufficiency bred out of them. They are less interested and able to find food in the woods, orchard, or lawn to supplement the ration you serve them. Some of these birds can be said to be more anxious than the Cornish-Rock Cross, but in my experience that means they get out from under the car when you are backing out of the driveway. Temperament differs more decidedly from one individual chicken to another—up and down the pecking order, for instance—than it does from one breed to another. E. B. White advised the beginning poultryman to "keep Rocks if you are a nervous man, Reds if you are a quiet one," but it would be a mistake to avoid Reds for fear of being surrounded by a flock of neurotic birds. They aren't.

A WORD ABOUT ROOSTERS

YOU DON'T NEED ROOSTERS TO GET EGGS, BUT YOU DO need roosters to get chicks. If you decide you want your layers to raise chicks (or just want this option, as fertilized eggs are perfectly fine to eat—more on this in Chapter 7), you'll need one rooster for every ten to twenty hens. If you have more than one, they will fight until they establish a pecking order and then will generally stop doing damage to one another. You don't need to be afraid of them—they are not usually given to attacking people—but you can certainly choose the ones to keep based on their temperament.

Roosters add to the interest and complexity of your barnyard whether you want fertile eggs or not. Colorful and dramatic, they will also help protect the hens against predators. They do, however, make a lot of noise, and whether this is seen as a positive or negative attribute is another matter of taste. Contrary to the message conveyed by Saturday morning cartoons, roosters crow at any and all hours of the day, not just in the morning, and it will be worth consulting with any close neighbors before subjecting them to what for many is a pleasant ringing in the air that epitomizes country life.

This ringing is now one of the legends surrounding my wedding. One of my earliest get-rich-quick-with-chickens schemes had to do with growing cocks for their hackle feathers which, I had heard, if sold to a fly fisherman for fly tying, were worth several times what a roasting chicken was worth. With youthful anticipation I filled the wood shed/chicken coop with fifty cockerels. Like every chicken, they started out very cute and fuzzy, and we thought no more about it than that. A week before I was to be married in the backyard—immediately adjacent to the coop—they all found their voices at once and began crowing. The combined decibels of fifty crowing cocks made it impossible to carry on a conversation inside the coop, and it wasn't much better outside. In desperation, the day before the wedding I killed the five loudest cocks, and they all fell silent for the next 48 hours while they reestablished the pecking order. The wedding went off perfectly—except for the rain, but that is another story.

HOW MANY GROWN CHICKENS DO YOU NEED?

You want to have more than just one chicken because they are social creatures and—like you and I—won't thrive without companionship of some of their own species. Whether you have a rooster or not is another question (see sidebar), but two laying hens with or without a rooster, or two meat birds, is the minimum. The number you choose beyond two will depend mostly on how much product you want in the end.

When it comes to meat, the questions are: How much freezer space do you have? How many chickens do you wish to roast, fry, stew, cordon bleu, or fricassee and serve up to your family in a given period of time? How often does your family eat chicken? Every other week? This means that twenty-five chickens is a good round number for the average family to attempt, if you have freezer space for them. Plan on a long day at the end with two or three helpers to butcher all your chickens (see Chapter 8), or find someone locally who will do it for you.

If you are raising laying hens, you'll want to consider how much you like omelets and how much refrigerator space you have. Because you keep layers longer than meat birds, and because their production is not consistent year-round, that figuring is more complicated. At the height of the summer laying season, each hen will lay an egg each day, and if you have not saved up a supply of used egg cartons to package them in and cart them off to work to give or sell to your enthusiastic coworkers, the eggs will displace much of the contents of an average-sized refrigerator before fall. When the chickens begin to molt and the weather cools into the winter, production will slow to a level you can consume and may fall below that during winter, when the less hardy breeds may stop laying altogether.

In summer three hens will provide a family of four with about a dozen and a half eggs each week, which amounts to roughly the current per capita

consumption in the United States. To carry this family through the winter and provide a few eggs for the neighbors in summer, you should add a couple more hens, and because marauders can strike without warning, you should add another as a spare. For a normal family, then, half a dozen layers is an appropriate and very manageable number.

These figures are for adult birds—you'll need to take mortality rates of chicks into account if you decide to acquire your chickens at an earlier stage of life. A hatchery won't send fewer than twenty-five chicks because they need to keep each other warm during the journey to your post office. If you ask them to send half laying hens and half meat birds, you can plan on putting ten or twelve broilers in the freezer in a couple months and having a lot of eggs in about 6 months. Decide ahead of time what you will do with the surplus eggs (see Chapter 7 on storage options), and start keeping your used egg cartons now. Consider asking friends to hoard theirs for you as well.

Which Comes First? And Where Can You Get It?

Once you decide what kind of chickens you want and how many you need, you'll need to figure out at what stage of your chickens' lives you'd like to get involved. Will you start with eggs for incubating, day-old chicks, pullets (female chickens less than a year old) that are about to start to lay eggs, or hens a year or more old? Each choice has its advantages and disadvantages, including cost and convenience, which I'll briefly discuss here. Specific instructions on incubating eggs, raising chicks, and selecting adult birds are found in later chapters.

A stack of egg cartons are ready to be filled in Tom Powers' chicken house. Powers said he sees the eggs produced by his chickens as a luxury food, because their freshness can't be matched by store-bought eggs.

Incubating fertilized eggs is somewhat tricky—a first-time chicken farmer can expect

Above: A bantam hen relaxing on Carrie Maynard's property. According to Maynard, getting some sun and a nap is a part of every chicken's daily routine.

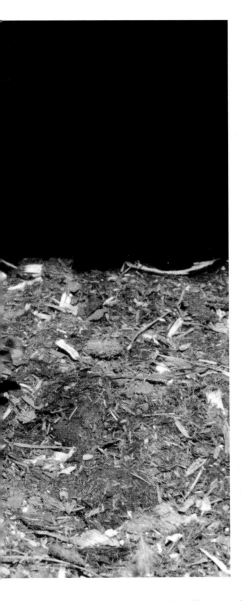

only about 50 percent of his eggs to survive into adulthood. However, it can be richly rewarding to know your birds from the time they break the shell open and crawl out of it soggy and hungry—not to mention cheaper than buying grown chickens. You can buy fertile eggs for hatching from a mail-order or more local hatchery. This method of raising chickens requires the most equipment. You will spend more than $150.00 for a small incubator with an automatic egg turner. Still, the younger you start your birds, the larger and more profound your experience of chickendom—and life on earth in general—will be.

If you want to have the excitement of fluffy chicks without the efforts of incubating eggs, you can buy them from a hatchery via U.S. mail, find a local hatchery, or buy them in the spring from your local feed store. This method is cheaper per bird than buying grown chickens, and it eliminates the big question of whether the eggs will hatch. But growing chickens to maturity is still a hazardous business—only 70 to 100 percent of chicks survive to adulthood.

Starting with pullets or adults will allow you to ease into the whole chicken-raising business slowly. You may be surprised at how many chickens are living within a short distance from your home.

CHAPTER TWO
Where Your Chickens Will Live

"*The ancestors of the domestic white-egg varieties were jungle dwellers, and sought safety and rest on the high limb of a tree or in the seclusion of the underbrush. . . . The housing of poultry is necessary, however, for there has been a general enfeeblement of most poultry stocks, which is the penalty we have to pay for breeding under unnatural conditions, for in-breeding to fix desired characters, and for using in the breeding flocks specimens which, though they show the racial type to the highest degree, are often the least fit in terms of bodily vigor and vitality. Artificial protection is necessary to offset this general enfeeblement.*"

— LESLIE E. CARD AND WILLIAM ADAMS LIPPINCOTT, Poultry Production, 1946

You will want to provide a home for your chickens for two primary reasons: (1) to keep animals that would eat your chickens (before you do) away from your birds, and (2) to moderate the climate for them, as extremes of heat and cold can make your chickens sick and/or miserable and inhibit maximum production of meat or eggs.

How you accomplish this task depends as much on your own ingenuity as on universally accepted prin-

Opposite: One hen finds a hole in the fence to her liking and gets out to stretch her legs at Tom Powers' home. Powers patched up the fence because the chickens were helping themselves to produce in the garden before the humans could get their share.

Above: Tom Powers' rooster stares down a visitor to the chicken house. Most of Powers' flock of 15 birds were taken by a coyote, he believes, one late fall morning. He wasn't home at the time, but in the past, "when a predator strikes, the cries of fear you hear from the chickens are heart-wrenching," he said.

ciples of poultry housing. I had a neighbor who let his birds have the run of an abandoned car in a pasture. The car had no glass in its windows and no backseats. The birds went in and out through the windows and laid their eggs in the trunk, where it was dark and they had some privacy. My neighbor opened up the trunk to get the eggs out. Another chicken farmer friend decided that what he wanted to look at every day out his kitchen window was an octagon-shaped hen house—a gazebo with a little cupola on top. He soon regretted this plan when he realized how much time and effort was involved in cutting, framing, and siding such strange angles and how difficult it was to get the pieces to match up correctly at each of the eight corners. Yet another keeper I know was visiting Peru some years ago and was intrigued by a small stone structure about waist high, with a hole in the front and large stone rolled to the side. At night when the chickens were inside, the free stone was rolled in front of the hole to seal out predators.

INITIAL DESIGN CONSIDERATIONS

Every chicken coop needs to have access to light and air, a way to keep the chickens in and the predators out, and a roof to protect the birds from inclement weather. In addition to providing a home for the chickens, the coop should be a place that you can tolerate spending time in, because you'll be going inside to feed, water, collect eggs, and, occasionally, clean. Ideally, you'll be able to walk in and shut the door without crouching or stooping.

Keep open the possibility of having separate compartments in the coop, too. You'll probably wish for this one day for any of a number of reasons, including opportunities to house chickens of different ages apart, give brooding hens a private spot to sit on their eggs, and allow the start of a batch of meat birds where the layers can't step on them. If you split your coop in two, you increase your options manyfold.

Before you begin breaking ground for your new chicken coop, take a moment to think about what makes the most sense for your climate, land, and chosen chickens. If you live someplace very temperate, you won't have to worry about keeping the house warm in winter, although you may need to make sure there's enough airflow that it won't get too hot. If you have lots of land on which the chickens will roam, you'll need less space inside your chicken house. Are you planning to raise layers in a northern climate, where they will require light even when it's dark outside? If so, you'll probably need to install electricity. Will you have to lug a bucket of water from your kitchen to the site, or do you want to be sure that a garden hose reaches?

These questions and more will come up again as you begin to envision your chickens' home-to-be, and I recommend that you pay attention to them. Keep common sense in mind—and remember the cupola.

THE LAND

LOCATION

Where should your chicken coop be? Not so far that you won't enjoy walking there, but not so close that you feel as though you are living in the coop yourself. If you put it in a spot that you pass by often, it will be easy to notice if things look awry. If it snows much where you are, think about how much shoveling out you will have to do to get access to it in February. It should be oriented so that its windows face in a southerly direction to take advantage of the warming and drying of the sun. If you have a fenced-in run, the run needs to be on the same side so that it dries quickly and doesn't remain a mud hole for long periods of time. For that matter, you don't want to locate the coop in a swamp, either. Put it somewhere that is reasonably well drained and where local flash floods or melting snow and ice won't run into it.

The diagram below shows how to take slope and orientation into account to get airflow, water drainage, and the warming and drying effect of the sun. Partway up a south-facing slope is ideal.

POOR CHOICE:
THE TOP OF THE HILL IS TOO EXPOSED.

POOR CHOICE:
IN A DEPRESSION, THE DRAINAGE AND AIRFLOW ARE INSUFFICIENT

IDEAL CHOICE:
A SOUTH-FACING SLOPE PROVIDES GOOD LIGHT, DRAINAGE, AND AIRFLOW.

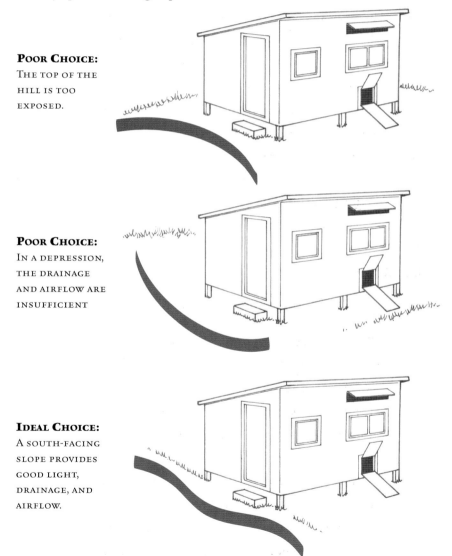

SLOPE AND ORIENTATION OF THE CHICKEN COOP SITE

"Being proud of a farming enterprise is easy if the farm and related buildings are kept looking nice and orderly. . . . Pride of ownership is important to the self-concept of each poultry farmer. Proper landscaping and regular maintenance of buildings and equipment make the farmstead look much better and make it easy for the owner to show pride in his farm."

—Oklahoma State University
Extension Facts F-8210,
October 1994

SPACE

Recommendations vary—depending on whom you talk with or what you read—for how much space you need to allow for your coop. Your chickens will need more room if they are kept inside all the time and less if they have access to the outdoors. The common rules of thumb range from 2 to 10 square feet per bird, perhaps with allowances made for the relative size of the birds. That means that a chicken coop for a dozen birds should be between 24 and 120 square feet—something between 4 feet by 3 feet and 10 feet by 12 feet. (As you can see, there are no

 Above: Frye the piglet readies himself for bed with a flock of meat chickens at Lianne Thomashow's home. The birds curl up with the pig for the night for warmth. The hog also has been a guard dog for Thomashow's chickens—she hasn't had a problem with predators since getting the piglet at the Fryeburg Fair in Fryeburg, Maine.

hard-and-fast rules.) The large industrial producers, of course, allow less space, trying to increase the return on investment in buildings and cages, or trying to keep them still so that more energy is converted to meat or eggs and less to running around pecking and scratching.

My own experience has taught me that too much room is better than too little. I started with two dozen layers in about an 8-by-8 foot area, which was a bit crowded even before a human entered to gather eggs. After Mr. Raccoon carried off half the chickens, I discovered that the remaining hens, their eggs, and my shoes, remained clean and the coop remained a relatively pleasant-smelling place. Because my coop was less cramped, the litter stayed dry and comfortable for the hens, and the coop was both pleasant to be around and easy to clean out for me.

RECESS FOR THE BIRDS

If you have the room, I would encourage you to let your birds roam as free as possible, within reason. Chickens that can spend time outside need less floor space inside. Keeping chickens on range, which is to say in a field of grass, can benefit both the chickens and the grass if they don't stay in one patch too long. You can create a shelter on wheels so you can move it every day or so to a new patch. You can either fence them in with portable electric fencing or let them go, knowing that chickens don't tend to wander too far from the coop. Grass farmers make a low flat pen that keeps the birds confined in an area limited to a few square feet per bird in order to precisely control where they graze and drop their manure to the maximum benefit of the grass. You can also build a run if you are short on space, or if you want to prevent the chickens from wandering off to sample your garden.

Above: A Columbian Wyandotte hen finds a good perch on a small log behind the chicken coop.

THE STRUCTURE OF THE COOP

THE WALLS

Most chicken coops are made using wood-frame construction like most houses and other small buildings. Lumber is generally available and relatively easy to cut and nail together. You will build vertical 2-by-4 stud walls with some kind of sheathing, either planks or plywood, on the outside. One suggested plan appears below. Insulation in your coop is not necessary if it is tightly constructed, but plywood or other sheathing on the inside can create an insulating dead-air space inside the wall. Another advantage to an inside wall comes when it is time to shovel out the manure. I still have vertical 2-by-4 studs exposed on the inside walls of my coop, and so have to work my shovel around them. An inside wall—say of plywood, even if only a few feet high all the way around—would make it easier to scrape the floor clean.

THE FLOOR

There are basically three choices for the floor: dirt, wood, and concrete. You will find many experienced chicken people that will swear by each of these options, passionately preferring one over the other. One will most likely appeal to you more than the others.

Dirt

A dirt floor is cheap and easy to do, and has worked for many chicken farmers in many cultures. Simply build on your ground—if there's grass there, it will soon turn to dirt. It is harder to shovel chicken manure off of it, however, and it will turn to mud if the soil underneath is not gravelly or sandy enough to drain well. You can get a sense of this by digging a small hole on your chosen site and adding water.

Wood

Any old boards 1 or 2 inches thick, in widths as narrow or wide as you can buy at a lumber yard or scrounge from the dump, will make an acceptable floor for your coop. Underneath your planks you will need supporting joists, usually about 2-by-6, which themselves rest on something at the corners, whether cinder blocks, stones, or heavy posts sunk a couple of feet into the gound. A wood floor keeps your birds off the ground and is relatively inexpensive, but it will rot eventually.

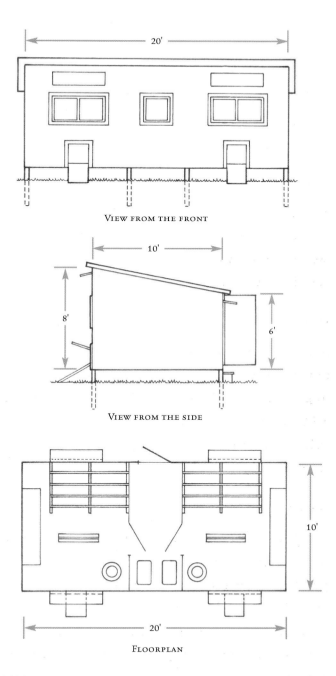

VIEW FROM THE FRONT

VIEW FROM THE SIDE

FLOORPLAN

THIS COOP COULD BE SPLIT IN TWO TO HOUSE CHICKENS OF DIFFERENT AGES

KEEPING
PREDATORS OUT

ALL OF YOUR OPENINGS—THE WINDOWS THAT OPEN as well as the vents near the tops of the walls—need to be covered with wire that will keep out the local predators. Weasels and mink will easily crawl through the holes in 1-inch chicken wire, and a raccoon in my neighborhood once grabbed a young chicken weighing a pound or two and pulled him, one bite at a time, through the 1-inch holes. Now I use galvanized half-inch screen (or hardware cloth, as it is called at the hardware store), which not only has smaller holes than chicken wire but is stiffer and won't bend as easily. The same raccoon pulled so hard on the wire while pulling on that dead chicken that he pulled out two of the ⅝-inch Arrow T-50 staples I had used to attach the wire to the post. Now I use ½-inch real wood staples that I drive in with a hammer instead of a household staple gun.

Even with these precautions, raccoons might get in. Any poultryman will tell you that raccoons are the smartest animals on earth after dolphins, and although they can't fit through the kinds of openings that a weasel can, they are dogged, clever, and patient. An opening they don't find the first year they may find the second or third. It took 3 years for one raccoon to find an entrance in my makeshift coop in the corner of the woodshed. I had stapled chicken wire all over the wall of the woodshed, but left an opening where there was a shuttered window. The window was closed, but it didn't matter when the raccoon went up into the loft above, found a space between the floorboards and the wall, crawled down inside the stud wall, and arrived in the chicken coop through the opening in the wire inside that closed window.

Concrete

Concrete is lovely to clean, impervious to rodents, and more or less permanent (which may be intimidating if you're not ready for a long-term commitment to chicken farming). Concrete, however, is the most expensive choice and requires the most effort to construct.

DOORS

You want separate doors that allow both you and your birds to get in and out of the coop. If you build separate compartments into the coop, you will want a separate door for the chickens to come and go from each compartment. Your chicken door need only be a foot or so high and a foot wide to allow your birds to pass through easily. If it is more than a few inches off the ground, you should build a ramp for your chickens to walk up and down. Make the ramp as wide as your door, and nail a piece of molding or strapping across it every 6 inches to give the birds plenty of traction. All the doors to your coop—and the windows too, for that matter—need to have a latch or other mechanism that can be secured at night against raccoons and other predators. Raccoons in particular are clever and agile and can easily open a simple hook and eye, for instance. You want a latch that closes—one that you imagine a toddler wouldn't be able to open.

To make life easier when it comes time to shovel out the litter, you may want to have an additional door just for that purpose. One coop I had was built on a slope so that the back end was about 2 feet off the ground. I put a wide door at floor level that was hinged at the top. It was fully as wide as the coop was: about 6 feet. When it was time to clean, I could prop it open with a stick and push the litter right out and down into a wheelbarrow with a snow shovel.

Above: A chicken's activity is dependent upon light, making windows, shown here at the author's coop in Vershire, Vermont, important. The sun is also a source of heat in the cold months.

WINDOWS

Windows and other ventilation holes are necessary for air and light, for both you and your chickens. After all, you both have to be able to see and breathe when you are in the coop.

Proper ventilation is critical for the health of a chicken. Chickens give off moisture, heat, and carbon dioxide as they breathe, and more moisture and ammonia rises from their manure as it mixes with the litter. These chemicals, dust, and various airborne pathogens can harm the birds if they reach high concentrations—which occurs when there is not adequate air exchange. The best way to assure proper ventilation is to put a series of holes or slots about 6 inches in diameter or 6 inches wide across the top of the north and south walls of your coop. These will provide natural cross-ventilation when necessary, without drafts.

It's important to keep your birds cool in summer. Chickens don't sweat, and with their feathers and relatively small surface area, they are not equipped to cool themselves as easily as other domestic animals. They will begin to suffer when the air temperature gets to about 95 degrees. You can tell they are too warm when they begin to pant like a dog.

In short, it's better to err on the side of too much air than too little. The first coop I had was nothing more than some standard 1-inch chicken wire stapled to a couple of upright 2-by-4s in the northeast corner of an old woodshed. The doors and windows of the shed had long since been removed, so ventilation was not a problem. The wind and even a fair bit of snow blew in, depending on the season.

Make windows that can open completely for warmer seasons, although be careful as to how they stay open—flat surfaces can prove disastrous. Wherever a chicken can sit, it will, and it will poop there as well. If windows tilt open, make sure they lie vertically up or down when they are open. I saw a coop once with windows that were hinged on the bottom and had a chain attached to the ceiling on top. The chain was just long enough to hold each window parallel to the floor. They opened into the coop like an old ironing board might have dropped down from the pantry wall or the way a changing table opens in a public bathroom. They provided a series of tabletops for the birds to sit on all summer, and they were a real mess.

In the winter, unless it is bitterly cold, you still should leave the south-facing air holes open. Don't be afraid to rely on common sense. If it smells nasty in there, your chickens probably need more air, and perhaps more litter as well.

Windows provide more than air—direct sunlight in the coop helps fight bacteria and keep the coop dry and even warm. If you have a cold winter where you are and the birds are inside for days or weeks at a time, they will be a bit toastier and much happier if they have windows in the south wall of the coop. If your climate is already toasty, put more windows in the other walls, and fix them so they can stay wide open all summer. If you live in northern climes, avoid putting more holes than you need to in the north wall, as arctic air can squeeze through in winter.

INTERIOR DESIGN

THE ROOST

Chickens are birds, after all, and birds are made to sit in trees. Not only do they feel safer, but they can modulate their body temperature better on a roost, particularly in winter when cool air settles. The chicken can fluff up her feathers to trap more body heat and cover her toes. Something as simple as a couple of old step ladders leaned up against the inside of the coop will create a number of roosts. You can also easily nail tree branches to strips of scrap wood. The tree

THE ROOST

Above: Hens and roosters in the chicken house. Their roost is made from a slender branch—decidedly rustic, but it works just fine.

branch should be about 1½ inches thick for larger breeds, 1 inch for bantams. Allow 18 inches between the roost and the wall, and between parallel roosts. Allow about 10 linear inches of roost per bird.

When setting up the roosts, you should keep in mind that chickens spend many hours there and create a significant buildup of droppings under them. I suggest you design the roosts so that you can move them out of the way when you want to clean. If you use an old ladder—or fashion one yourself—attach it to the wall with hinges so that you can lift it up (and hook or tie it to the ceiling somehow) when you are ready to clean.

Some people place a pan covered with wire mesh underneath the roost so that droppings fall through the mesh and accumulate in the pan. This method allows you to take a significant portion of manure out of the chicken house before it fouls the litter or your chickens. You can keep your birds clean with less litter. Chickens spend many hours on the roost and create a significant buildup of droppings there. You can also arrange chicken wire on either side of the ladder/roost holder to keep the chickens from getting underneath, where the droppings are heaviest. The pan and grate need to be cleaned every week or at least every month, depending on how many chickens you have. Keep in mind that any system that doesn't allow the birds to mix the manure in with their bedding is a system that requires you to handle unadulterated manure on a regular basis. It also might create more smell and attract more flies.

LITTER MANAGEMENT

I think it is more effective to allow more rather than less floor space per bird, lay in 5 to 10 inches of litter on the floor of the coop, and let the birds have access to all of it. They will stir their own manure into the litter with their scratching, particularly if you throw them some cracked corn or other grain occasionally. Add more litter as it seems necessary. With the right ratio of chickens to litter, the manure will virtually disappear and over time even begin to compost right underneath the chickens. With a thick soft layer of bedding, their feet and legs suffer less strain, and the slowly composting litter will even produce a little heat if left alone all winter.

NESTING BOXES

Hens like to lay eggs in a clean, safe, private place, but they will put them just about anywhere in the coop, if you keep them confined. If you'd rather not root around the floor searching for the eggs in damp (or worse) wood shavings or

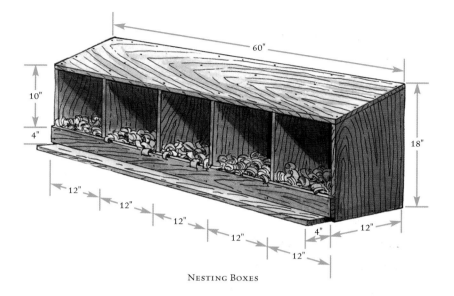

NESTING BOXES

chopped straw or whatever it is that you put between the floor and your birds, I suggest you build—or find, or buy—a nesting box. A nesting box is simply a box, a little bigger than a chicken, open on the front, that gives the hen a comfortable place to lay, with the added bonus that you both know where to find the eggs. And you may find more than you expect, as the hens will share these boxes. You need just one box for every four to five birds. They only use them once a day to lay an egg, and they are not territorial about them.

The general requirements are that a box be about a foot deep and a foot wide, with an opening about a foot high in front. The exact dimensions are not critical, but it should have a steep-pitched roof (try 45 degrees) so the birds don't sit on top of it and mess it up. Some people hang a piece of burlap down over the front of the box to protect the birds' modesty, as if the nesting box were a changing room in a clothing outlet store. To keep the eggs clean and the birds comfortable, you want to keep litter in the boxes, which is easier if there is a lip about 4 inches high across the bottom of the opening in front of the box.

Above: A pair of chickens—a Black Australorp, at left, and a Rhode Island Red hen—check out a visitor to their stomping grounds..

The main point of the nesting box is to provide some peace and quiet for the layers. A hen we had once had ignored a more open nesting box we had made for it in the coop and took to laying its eggs in a kind of den a skunk had dug out several years before under the front porch. It was nothing but a hollow in the gravel, but it was way back in the corner against the house foundation under the floor boards. It was private and protected. One neighbor of mine keeps fifteen Araucana layers that all use the same makeshift pair of nesting boxes he made by taking a couple of boards off the front of an old butter churn and dividing it down the center with another board. This contraption did provide roughly a square foot of floor space for the birds to nest in, but it was at least twice as tall as it was wide, and his birds didn't seem to mind. As long as the hen isn't crowded and she can get in easily, almost any size will do.

Finally, there are additions to the boxes that are entirely personal preference. Putting a fake egg into each box may help convince the hen that it is a good place to lay. I have a few plastic hollow eggs from one Easter that seem to work fine. I filled them with sand so they have something like the heft of a real egg. As for other nesting box amenities, Ulisse Aldrovandi, a famous Renaissance

"*There is no electricity at our barn, no running water, and no water heaters . . . No pipes to freeze, no motors to burn out. Without "labor saving" technological gadgets to help me, I save a lot of time by not having to fix them.*"

—GENE LOGSDON, *The Contrary Farmer*, 1993

naturalist and chicken expert, wrote that some people in his day put a thin piece of iron, the heads of nails, and sprigs of laurel in their nests "because these seem to have great power in driving off bad luck." He goes on to comment that "remedies of this kind . . . indicate the excessive zeal of the ancients." Zeal perhaps, but I will take whatever luck I can get.

CHICKEN COOP AMENITIES

ELECTRICITY

You might consider running electricity to your coop to light a bulb, especially if you have a full-time job. For significant portions of the year, by the time you get home from work you will be visiting your chickens in the dark, and when you are carrying water or pouring grain, there is not another hand left over to hold a flashlight.

There are a few ways to do this. One coop I had was close enough to the house to run a 50-foot extension cord to it. The cord lay right on the driveway,

and the snow plow chopped it into several pieces every time it snowed. It is safer and more permanent to bury an appropriate wire under the ground and install a grounded outlet or two in your coop.

In areas that have a colder winter, electricity is an even bigger

Above: Tom Powers' flock of chickens parade from their house when they hear the voice of their owner. Powers built the house in 1983, based upon an article he read in the magazine *Country Journal*. Using doors and windows found on the family farm, he followed the magazine's recommendation and has a separate entrance inside the building to the chickens' pen

boon, for a couple of reasons. One is that you can plug in a device that will keep the birds' water from freezing (more on that in Chapter 6). Another is that the number of eggs a hen lays is affected by the number of hours of daylight it is exposed to. In winter, there are fewer than 14 hours of daylight, and a hen may stop laying altogether. By putting a light on a timer and adding artificially to the hours of light your laying hens are exposed to, you can keep them producing eggs in the winter when the daylight hours are short. (See Chapter 7 for more on light requirements).

The Chicken Run

Chickens are healthier and happier outside, but you may want to take some precautions before allowing them to roam free, particularly if you are a gardener. The birds will eat and scratch up whatever you have planted, and will make large holes in which to cover themselves with dust. (Don't discourage this—give your chickens access to dirt, because their "dusting" activities discourage external parasites.) Consider creating a coop that opens onto two separate runs that can be alternatively closed off. This way the birds can have access to one run while the other recovers—over the course of anywhere from a month to a summer—from the abuse of a dozen scratching and dusting chickens. Make your fenced-in run out of chicken wire, which the feed or hardware store might call 1-inch poultry netting. It has holes that are octagon shaped and measure about 1 inch across. Get the galvanized version so that it will rot more slowly. You may also find a product called aviary netting, which is the same thing but with smaller holes. Wrap this around the bottom foot of your fence if you have small chicks.

Opposite: Lianne Thomashow built a coop in her barn in two days out of doors, windows and shutters found on their property. When she added more chickens to her flock in the summer, she raised the floor off the ground and enlarged the coop.

They can easily walk through the holes in the standard 1-inch chicken wire.

A four-foot fence is sufficient to keep the birds in, but you may want to make it tall enough for you to walk in. This is because it is well to put a chicken-wire roof on your run, both to prevent your birds from flying off (if you have bantams or other light birds that retain their ability to fly) and to make sure that a local owl or hawk won't swoop in for din-

Above: A strong ring of chicken wire around the coop and run is essential to keep predators at bay.

ner. At the bottom of the fence, consider burying the wire 6 inches deep and making a 90-degree angle underground so that the fence bends toward the out-side of the run. If you do this, a burrow-ing dog or fox will dig down along the wire until it runs into fence at the bot-tom as well. It will probably give up at this point.

Opposite: Carrie Maynard's coop was built to be expandable and portable. Originally built with dimensions of eight by eight feet, it is now twelve by eight with a roof that slopes from seven and one-half to six feet, and is the home for 38 chickens.

CHAPTER THREE

Getting Eggs to Hatch

"*There have been, indeed, suggestions as to why certain actions and reactions follow one another in orderly sequence. But with all the facts made known through patient and persevering research, and the penetrating minds which have pondered them, comparatively little progress has been made toward a knowledge of the how and why of this eternal mystery.*"

— LESLIE E. CARD AND WILLIAM ADAMS LIPPINCOTT,
Poultry Production, 1946

Once you have chosen what kind of chickens to keep and prepared a place to keep them, you are ready for the birds themselves. If you have decided to start your flock from fertilized eggs for the first time, be easy on yourself. The process is somewhat painstaking, and you must define success broadly because at least half the fertilized eggs you incubate will not hatch. Still, watch-ing a chick kick away the last pieces of shell and lay in the incubator exhausted after its struggle to arrive in this world, is one of the big reasons to have chick-ens in the first place, and makes any success worth your while.

Opposite: One of Tom Powers' hens strikes a statuesque pose for a visitor to the coop.

WHERE TO GET EGGS

Your choices for getting eggs are to order fertile eggs from a hatchery for anywhere from $1.00 to $3.00 each, or to collect them yourself from your own hens (or a neighbor's). If you are ordering them, the hatchery will probably make good on any eggs that arrive broken, but no one can guarantee how many will actually hatch. Fifty percent is a conservative estimate, and three-quarters is a good showing. Take this into account when deciding how many to order.

If you want to incubate and hatch your own eggs, you must first, of course, have a rooster. Although hens will lay eggs for a lifetime without ever laying eyes on a rooster, those eggs will not hatch unless they have been fertilized. The rooster accomplishes that. In order to service your hens adequately, the rule of thumb is that you will need one rooster for every ten hens. As your rooster grows older, which is to say after the first couple of years, he will slow down and may only be able to handle five hens.

HANDLING FERTILIZED EGGS

A typical cycle at the height of the laying season in spring and summer calls for a hen to lay an egg about every 25 hours, or an hour later each day, until the lay-

ing time starts to run
into the evening hours.
The hen won't lay in the
dark, so this usually

Above: A nest egg encourages hens to go to the nest for egg-laying. Carrie Maynard found wooden eggs on clearance at a home furnishings store, but she said any round object would work, including golf balls.

means about one egg a day for 12-15 days, although it depends on the season (less daylight equals less eggs). The rest period is merely overnight if you're collecting the eggs, but it's just about three weeks if the hen is brooding her eggs.

Although no one really knows why a hen lays every 25 hours instead of every 24, or some other round number, it has been speculated that nature invented this system so the hen would accumulate just the right number of eggs for her clutch before breaking her laying rhythm. During that week or two that the eggs are not kept warm, development of the embryo inside stops or slows to almost zero. As soon as the egg is warmed, it begins to develop again quickly.

If you incubate them yourself (and even if you put them in a nest for one of your hens to sit on), your role as chicken husband is to remove the eggs from the laying box as soon as possible after they are laid in order to keep them clean and safe until you have enough to fill your incubator. You should store fertile eggs at between 40 and 60 degrees in the open air, dry, for at least 12 hours but no more than 2 weeks—and preferably within 6 days—before incubating them. Store eggs large end straight up in a clean egg carton before they go into the incubator and at a 30-degree angle in the incubator.

Keeping the eggs clean is a major concern. Hard as they are, eggshells are porous. A living, breathing chick will grow inside, so the shell is made to allow oxygen in and carbon dioxide out. Even the natural oils on your skin may plug up the holes in the eggshell and inhibit the movement of these gases. For this reason, you should wash your hands with soap and water any time you get ready to handle the eggs.

If there is manure on the eggs, it is best to brush it off as well as you can without using any water or soap. An egg has built-in defenses against bacteria, but if you scrub with water, you can easily push tiny organisms into the egg through its minute pores.

BROODING BY HEN

You can let a hen incubate the eggs for you, or you can incubate them yourself in a mechanical incubator. One reason to let the hen do it is to save you the

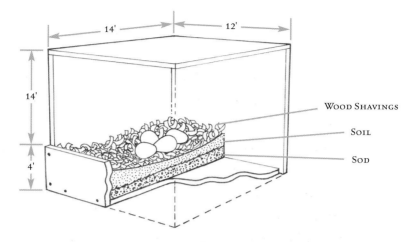

NESTING BOX FOR A SETTING HEN

time, effort, and constant worry it takes to reproduce the atmospheric conditions that prevail underneath a setting hen. Many breeds have had the inclination to brood bred out of them, however, because egg production stops while the hen sets. You may or may not have a hen that is willing.

If you choose to let a hen hatch the eggs for you, assemble a dozen or so eggs in a nest and see if any of your hens seems interested in setting. If one sets—if she stays where she is and gets mad when you try to take an egg out from underneath her instead of running away—you've found your broody hen. Set her and her eggs up in a darkened but well-ventilated nesting box in a place where the other hens can't get to it. While she is setting, she could be abused by her coop-mates or have her nest taken over by another. Also, when the chicks are eventually hatched, they might be eaten or trampled by the rest of the adult flock if they are allowed to mix during the first few days.

Because parasites can multiply and cause havoc with a setting hen, you may want to use cedar shavings for litter in the nesting box and consider sprinkling or spraying her with a chemical parasite control product available from a poultry supply operation or feed store. After that, she is on her own. Make sure she

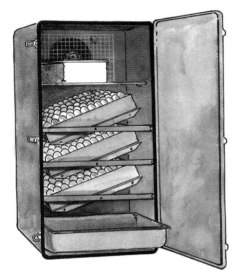

CHEST-TYPE INCUBATOR

TABLETOP INCUBATOR

can get to feed and water easily, and have the same ready for the chicks when they arrive.

ARTIFICIAL INCUBATION

Incubators are essentially boxes that help keep the eggs warm and moist—mimicking the environment under a brooding hen. They come in a wide variety of shapes and sizes, but for the backyard keeper there are two general approaches, depending on the scale of your operation. One is an incubator that sits on a table, and the other is a chest-type incubator that sits on the floor. The smallest tabletop incubator made for school projects might hold just two eggs, whereas the chest type holds perhaps 300.

The other major distinction between incubators is still air versus forced air. Still-air incubators have ventilation holes in the bottom and top, and as the cooler fresh air pushes in through the bottom holes, the warmer air exits from the upper holes. The forced-air type has a fan that constantly circulates the air within the incubator and introduces fresh air at the same time. Forced air is more expensive, but it keeps the temperature more constant in the incubator—this is particularly handy if the incubator is in a room where the temperature fluctuates more than 5 degrees in either direction over the course of the day. It also tends to offer a better idea of how much humidity is present, because those levels are more constant, too.

Read the instructions that come with your incubator carefully and follow them to the letter. Particular recommendations vary from one model to the next, but the following basic principles apply to all incubators.

Parts of an Incubator

A common tabletop incubator often has a see-through plastic lid that allows you to watch what is happening. It has a rack to place the eggs on, a heating element, a thermal switch that turns the heating element on and off in order to keep the temperature constant, a place to pool water in the bottom in order to maintain the correct humidity, and holes that you can plug or unplug for ventilation. An automatic egg turner is a labor-saving extra you will want to seriously consider. Either you or the machine will have to turn the eggs at least three times a day, and the more times you turn, the better chance your chick has of hatching. A fan to convert the still-air type into a forced-air type is another extra.

Placement of the Incubator

There are a few rules that govern the placement of the incubator in order to avoid certain disaster. The first is that it should never get direct sunlight. Constant temperature is critical, and direct sun will kill off your embryos quickly by raising the temperature in the incubator. The incubator should be

kept in a 70- to 80-degree
room that is safe from pets,
small children, and other
predators.

THE KEYS TO SUCCESSFUL INCUBATION

It only takes 21 days for an
incubated chick to develop
inside its shell to the point of
hatching. During that time,
the temperature and humidity
must be just right, the egg
must have fresh air, and it
must be turned frequently to
keep the developing embryo
near the center of the egg.

PLACEMENT OF THE EGGS

Incubating eggs should lie on
their sides as they would on a
flat surface, with the larger end
slightly elevated. The racks that

come with most incubators take care of this. The developing chick's head will
orient toward whichever end of the egg is higher. Not only is there more room
in the large end, but there is an air space in the large end of each egg that the
hatching chick needs while it is hatching, as described later in the chapter.

 Above: One of Lianne Thomashow's hens takes a look at the hay set aside for the horses in the barn shared with the chickens.

Above: A Columbian Wyandotte hen settles down for egg-laying. The hen's back feathers are worn due to breeding by the rooster.

TURNING

Turning the eggs several times a day, up to day 18, is critical to the developing embryo. The embryo grows on top of the yolk, with surrounding layers of egg white serving both as nutrition and protection while it is most fragile. If the egg is not turned, the embryo will slowly float up though the white and come in contact with the shell membrane. If this happens, it is likely that it will get stuck there and die. For the last 3 days of incubation, the egg should not be turned at all. At this point, the chick is positioning itself to break out, and turning will hamper it.

A hen on her nest turns her eggs every 15 minutes, either by shifting herself to get more comfortable or by reaching under herself with her beak to roll them. If you are turning your eggs by hand, you only need to turn them three times a day, as long as the first and last times are at the very beginning and very end of your day. What is important is that it be done at regular intervals and that the longer interval at night is as short as you can reasonably make it.

In order to keep track of which eggs you have turned, use a pencil to mark each egg with an X on one side and an O on the other. Pick up the front row of eggs, roll the rest gently in one direction until all the Xs or Os are at the top, and then put the removed row at the back. Be particularly gentle with the eggs, as the tiny, fragile blood vessels of the developing embryo can break with rough handling. If you can turn them more than three times per day, up to eight or ten, it might improve the chances of some of the eggs. An automatic egg turner can be added to a typical home tabletop incubator for about $50.00 and will do the turning for you.

TEMPERATURE

Different types of incubators require different temperatures, but it is impor- tant to understand that these temperature differences are quite small. A fluc- tuation of much more than half a degree either way can get you into trouble, which is why the sun shining on your incubator can quickly kill whatever is in those eggs. Also, the embryo can tolerate lower temperatures more easily than higher ones. Keep a thermometer inside that you can read through a window so you can monitor the air temperature in the incubator. Over time, chicks begin to generate their own heat, and you may have to adjust the temperature down. In any case, you should turn on your incubator a day before you put eggs in it to make sure the temperature is correct and holding steady.

In a still-air incubator, the air temperature at the top of the egg is as much as a degree or two warmer than at the bottom, and you need to measure the temperature of the air at the level where the embryo is developing. Because the embryo tends to float near to the top of the egg, the bulb of your thermometer should be about a quarter inch below the top of the shell when the shell is laying on its side.

HUMIDITY AND THE "AIR SPACE"

An eggshell is porous, and water constantly evaporates from it. Because of this, the humidity in the incubator needs to be correct throughout incubation or your chicks will be unable to hatch from their shells.

Soon after an egg is laid, an airspace forms in the large end—if you've ever peeled a hard-boiled egg, you've seen this. Over the period of incubation, as fluid in the egg evaporates through the shell, this airspace grows. When it comes time for the chick to hatch, it will poke its beak first through the membrane that separates his own compartment from the airspace, and then poke its beak through the shell in a circle to break off the cap of shell over the airspace. If too much fluid has evaporated from the egg and the chick is too dry, the chick could be dehydrated by the time it has to break out of its shell. What's more, as he pokes through the shell, it could stick to him and make it impossible for him to break free. If humidity is too high, however, the airspace at the large end might not be large enough, and the chick may begin to

"pip" or break through the shell in the fluids under the airspace. If this happens, the chick can drown before it gets the shell open.

Above: A pair of Dark Brahma hens who are molting enjoy the sun on a warm morning.

MEASURING HUMIDITY

The humidity in an incubator can be measured with a hygrometer, which is a thermometer that has a bit of water-soaked cloth wrapped around its bulb. Most forced-air incubators have one; if you're using a still-air incubator, you can buy one, but you'll probably want to use the "candling" method (discussed later in this chapter) to measure humidity. Most often the humidity is measured in "degrees Fahrenheit wet-bulb," which is the temperature recorded by the hygrometer as the water around its bulb evaporates. You can also talk about relative humidity in terms of a percentage, as the weather report does. By this measure, the humidity in your incubator needs to be at 60 percent until the chicks begin hatching, when it should increase to 70 percent. The equivalent wet-bulb measurement is 84 to 86 degrees for incubation and 90 degrees when the chicks start to hatch.

ADDING AND SUBTRACTING HUMIDITY

Early in the incubation period, it is easiest to increase the moisture content of the air in the incubator by plugging one or more of the bottom ventilation holes to restrict the airflow and therefore keep more moisture in. During the last few days of incubation, the need for moisture competes with the need for oxygen unless you increase the humidity some other way. You don't want to block ventilation at this point—the chick inside the egg needs more oxygen and needs to get rid of more carbon dioxide as it develops. An incubator without enough airflow will fill with too much carbon dioxide and the unhatched chicks will suffocate.

One way to solve this dilemma is to put a humidifier in the room where the incubator is. This way the fresh air going into the incubator is already more humid. Another way is to put a soaked sponge in the incubator. The humidity in the incubator depends on the amount of surface area of the water that's in it. The sponge has lots of surface area and will keep the air moist.

CANDLING

Candling an egg means to shine a bright light through it so you can see something of what is going on inside. You are looking for two things: the presence of

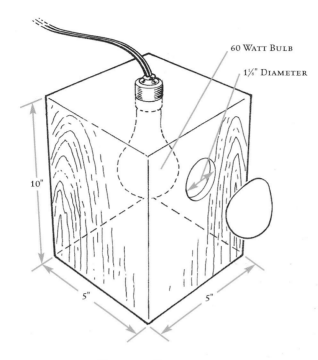

60 WATT BULB

1⅛" DIAMETER

10"

5"

5"

HOMEMADE CANDLER

a living embryo and the relative position of the airspace, which may indicate that the humidity is too high or too low.

In former times, candling was literally done with a candle, but a 75-watt lightbulb is easier to use. The feed stores and catalogs sell flashlightlike gadgets made for the purpose, but a homemade candling box is not difficult to build, as you can see from the diagram, above. Both the gadgets and the box allow the light through a hole about an inch in diameter so that more light goes through the egg and less blinds the candler. In either case, you can see better if you candle eggs in a darkened room.

Candling the eggs requires taking them out of the incubator and cooling the whole apparatus down, so you don't want to do it too often (or too slowly). It is best to candle the eggs at the end of the first week (to find infertile eggs or

"*G*aining an understanding of the developmental anatomy of the chick is one of the major hurdles that generations of biologists and medical students have faced. Something . . . is known about the chemical events associated with each stage, and how these are regulated by a constantly shifting pattern of gene activity. This knowledge is only a faint glimmer in a vast darkness, however, and the most sophisticated embryologist, observing a wet, scraggly chick laboriously peck an opening in its shell, feels a sense of wonder, and, at many times, surprise."

—CHARLES DANIEL AND PAGE SMITH,
The Chicken Book, 1975

those with dead embryos) and at the end of the second week (to remove any eggs that have died since the first candling). As hatching approaches at the end of the third week, it is best to leave the eggs alone and let them prepare to hatch, especially after day 18.

CHECKING FERTILITY

If the egg is not fertile, or if the embryo has died, you wan
rather than later so you can remove the egg from the incu
rotten eggs give off gases that you don't want your other e
eggs can also occasionally explode.

 Above: Held in front of the light of a candle, an egg is checked to see if it is fertile. If it is, the disc of a chick's embryo is visible. Fertile or not, the egg is edible.

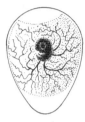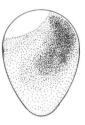

| LIVING EMBRYO AT ABOUT 7 DAYS | LIVING EMBRYO AT ABOUT 14 DAYS | LIVING EMBRYO AT ABOUT 14 DAYS | DEAD EMBRYO AT ABOUT 7 DAYS SHOWING THE "BLOOD RING" | DEAD EMBRYO AT ABOUT 14 DAYS |

When you candle an egg, you're looking for one of three basic configurations. What you want to see is a small spot with spiderlike legs—the embryo's blood vessels—branching out from the center. This is a living embryo. A uniformly opaque egg with a shadow cast by the yolk was never fertile and an egg with a darker shadow in the middle surrounded by a discernable blood ring has died (the blood has moved out away from the embryo).

Checking the Size of the Airspace

Recognizing an egg with an appropriately sized airspace comes with practice. At first, you'll need to refer to a diagram. If your eggs seem to have airspaces that are much smaller than the average, try to increase the humidity in your incubator by

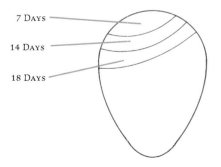

7 DAYS

14 DAYS

18 DAYS

THE SIZE OF THE AIRSPACE GROWS OVER TIME

adding a sponge or filling more of the compartments with water. If the airspaces seem too small, consider opening more vent holes to allow more air circulation.

HATCHING

From day 19 to day 21, when the chick comes out of the shell, it takes the yolk of the egg into its stomach through the navel. It hatches fully fed and has time to dry off and get its bearings before it has to find food in the outside world. On hatching day, the temperature can be reduced to 95 degrees, but the humidity should be increased to 70 percent or 90 degrees wet-bulb. If you have a see-through incubator top, it should be fogged up. Put a double layer of cheesecloth on top of the rack to make a cleaner, more comfortable place for the new chicks to spill onto and dry out. It is important to leave chicks in the incubator for 24 hours after they hatch, until they are completely dry and fluffy. Chicks taken out before they are dry will chill quickly and die soon after. Have your brooder setup ready to go with water, feed, and the heat lamp turned on, and remove the dry chicks to the brooder once a day for the day or two it takes for them all to hatch.

CHAPTER FOUR
Raising Chicks

"*Why in God's name does everyone want to go into the chicken business? Why has it become the common man's Holy Grail? Is it because most men's lives are shadowed by the fear of being fired—of not having enough money to buy food and shelter for their loved ones and the chicken business seems haloed with permanency? Or is it that chicken farming with each man his own boss offers relief for the employer-employee problems which harry so many people? . . . Again I repeat, why chickens? Why not narcissus bulbs, cabbage seed, greenhouses, rabbits, pigs, goats? All can be raised in the country by one man and present but half the risk of chickens.*"

—BETTY MACDONALD,
The Egg and I, 1953

Cute and fuzzy as they are they are, newborn chicks are also at their most vulnerable during the first few days and weeks of their life and need

Opposite: A set of the "Meat and Eggs" chicks from the hatchery graze from a chick feeder that keeps their food in place in the author's chicken coop.

particular care and attention for them to have the best chance at survival. If they are warm and dry, and have access to food and clean water, most of them will grow and prosper under your care.

GETTING READY: WHERE WILL YOU KEEP THEM?

Where you raise your chicks depends on the season, the number of chicks you want, and the relative tolerance of other members of the family for live-stock projects in the house. If the temperature outside during the day is roughly the same as the temperature inside the house, especially if it is spring and the trend is toward warmer weather, it is simplest to brood the chicks in the space where they will live when they are adults. This saves you having to move them as they grow, but it will only work if they have that space to themselves or if the chickens already in the coop can be kept entirely separate from the new chicks. The grown chickens will trample, eat, and otherwise terrorize the new ones.

Even in good weather, many small-scale operators prefer to keep their chicks in the kitchen in a cardboard box, at least for a few days. Brooding chicks in the house for the first week or so, particularly when the weather is cool or cold, helps you to worry less about drafts, and it allows you to watch them easily and often, both for entertainment and to monitor their well-being. Chicks grow very rapidly, however, especially meat birds, which go from a bit of fluff that fits inside an egg to 4-pound broiler in 8 weeks. You will need to keep switching boxes, or adding on in some fashion, to give them enough space. After a couple of weeks, they will have more than doubled in size and will start looking too big for the kitchen. The smell may also become an issue.

KEEPING YOUR CHICKS WARM

There are two parts to keeping your chicks warm: providing them with a heat source and protecting them from drafts.

THE HEAT SOURCE

The air temperature around the chick as it skitters from feed to water and back should be about 95 degrees Fahrenheit for the first week of its life and should decrease by 5 degrees per week after that until it reaches 70 degrees. At that point, chicks are able to keep themselves warm, unless it is particularly cold outside.

The easiest way to provide warmth is with a lightbulb. If you have one or two dozen birds and the weather is relatively mild, a regular 60- or 100-watt lightbulb will do. Start with the bulb about 18 inches off the floor, and raise it about 3 inches per week in order to steadily lower the air temperature around the birds. Use a porcelain socket because a plastic socket will melt after enduring even a 60-watt household bulb hung upside down and left on most of the day. You will also need to put the bulb in a reflector—one with protective wire over the front of it to reduce the chances of an accident that could break the bulb and, in the worst case, cause a fire. If the weather is cooler when you are starting, a 250-watt infrared heat lamp from the feed or hardware store will be necessary.

DRAFT PROTECTION

Draft protection for the first few days often takes the form of a corrugated cardboard box with sides approximately a foot high. For added protection, you should partially cover the top with another piece of cardboard. The birds are very small for the first few days and don't take up much space. Twenty-five chicks will do fine for a few days in a box about 18 inches wide and 2 feet long.

The box will prove too small pretty quickly, however. To provide adequate space that you can expand as needed, a circular ring of cardboard or other material called a draft guard or chick guard is the best solution. It should be about a foot high. Not only will the chick guard keep drafts out and tend to hold in the heat from your light bulb, but it will also prevent new chicks from wandering to far corners of the coop and forgetting where the feed, water, and warmth are. The chick guard has no corners, so the chicks will not pile up and smother each other in a corner, as they might do if frightened or cold. Make sure the circle is always big enough for them to get out from under the heat if they want to.

After a couple of weeks they will need half a square foot per bird, and in about 4 to 6 weeks—by the time their feathers grow in and they can keep themselves warm—they will need 1 square foot per bird.

One February in northern New England during a cold snap, as I was making one of my first attempts at chicken rearing, I put my chicks in a kid's plastic swimming pool in the basement. The sides were almost a foot high, it was round, and it kept them off the dirt floor. I put some litter in it and figured it would work just fine. The medical student friend I have mentioned before took a disused incubator from the dumpster at the hospital, a long-necked, lead-footed piece of equipment made to keep infant humans warm in the maternity ward. Eight out of twenty-five of my chicks died in the first 2 weeks. I now believe this was because the house was 150 years old and the stone foundation let in a lot of brutally cold air that washed invisibly over the sides and into the pool. The moral of the story: Pay attention to the chicks' surroundings.

You may suspend a thermometer from above the brooder apparatus so that it dangles about 2 inches above the floor if you like to have objective scientific proof of the temperature in that microclimate immediately surrounding the thermometer. It is best to hang it so that it doesn't disappear forever in the litter, and so you get the reading at the level where the chicks live, as the air above and below may be somewhat different.

Visual observation, however, rather than a thermometer, is the better way to achieve the correct temperature. Start with the lamp about 18 inches off the floor, and be aware of what the chicks do over the next hour as the temperature

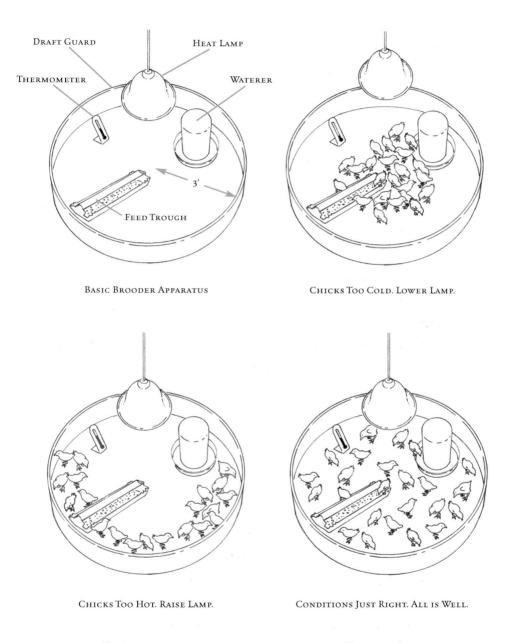

DRAFT GUARD

HEAT LAMP

THERMOMETER

WATERER

3'

FEED TROUGH

BASIC BROODER APPARATUS

CHICKS TOO COLD. LOWER LAMP.

CHICKS TOO HOT. RAISE LAMP.

CONDITIONS JUST RIGHT. ALL IS WELL.

THE ILLUSTRATION AT THE UPPER LEFT SHOWS A BASIC BROODING SETUP
INCLUDING FEEDER, WATERER, AND THERMOMETER SURROUNDED BY THE DRAFT GUARD,
WITH THE HEAT LAMP SUSPENDED FROM ABOVE.

Above: The author's chicks take a break in their brooding box.

in your setup stabilizes. If the chicks huddle under the lamp, they are too cold and the lamp is too far away. If they crowd the wall or arrange themselves outside what has become a circular hot spot directly underneath the lamp, it is too close. Move it up or down a couple of inches at a time until the chicks mill around all over the floor, comfortably eating and drinking or spreading themselves out more or less evenly across the floor to sleep.

Strictly speaking, chicks need to have a range of temperatures available, between about 75 and 95 degrees at first, so that they can choose a tempera-

ture comfortable to them. Not all of them are comfortable at exactly the same temperature at the same time, which is why you know you have it right when they spread themselves out. You also know you have it right if they are eating and cheeping pleasantly. If their peeping is loud and anxious, they are probably too cold or hungry.

This is not to say that a quiet brood of chicks is necessarily a bad thing. Chicks need a lot of sleep, and for much of the night and parts of the day, they trade peeping for sleeping. They rouse themselves quickly, however, when there is any activity nearby, or if they have a visitor, and go back to eating and drinking, which is what you want them to do.

KEEPING YOUR CHICKS HAPPY AND HEALTHY

To keep your chicks healthy, you need to provide air, litter, and light. A constant supply of fresh air is the most important way to keep the chicks' living space dry because the chicks' breath and manure constantly release water into the air. Keep in mind that chicks' production of manure by volume increases as fast as the chicks themselves. We have talked about the dangers of drafts. How are they to be warm, draft free, and have enough fresh air? It's a delicate balance, and it's easier when the air outside is warm. Even if this isn't the case, make sure the vents on the south side of the coop are open if the birds are outside, or that the windows are ajar if the chicks are in a small, tightly sealed room.

For litter, for the first 2 days—and only the first 2 days in order to avoid leg problems—put a layer of newspapers or paper towel on the floor inside the draft guard and spread their feed on it. This way they can find it easily using the peck-and-scratch method that all chickens are born knowing. After this introduction to pecking, you can spread pine shavings on the floor. The reason to wait are twofold: If you had spread crumbles in the litter initially, they would have disappeared quickly to the floor underneath it. And if you leave the newspaper there after the second day, the chicks begin to lose their footing, and those most susceptible to leg problems may begin to drag themselves rather than walk from here to there.

Natural light, of course, is a must, and whatever housing you have cobbled together should have it, at least after the kitchen stage is over. Direct sun on the floor of the pen dries it out and kills unwanted microorganisms. Just as important to newly hatched chicks is extra artificial light. Chicks grow better if they eat during the night for the first 2 days, and a light helps them to find the feeder and gives them encouragement to go find it in the first place. If your brooder is a lightbulb, you have satisfied the requirement. Make sure to turn it off for at least a half hour a day though, or your chickens may panic and pile up if the electricity goes out.

PROVIDING ENOUGH WATER

There are various sorts of feeders and waterers available for starting chicks, and you will need to have two or three sizes on hand to accommodate your fast-growing birds. They will double their intake of water in the first 2 weeks and double it again 2 weeks after that. For the brand-new chicks, you will need to spend $3.00 or so at the local feed store on a few galvanized metal bases that

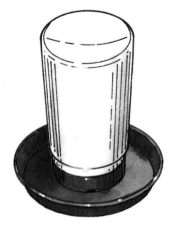

CHICK WATERER

screw onto the top (which becomes the bottom when you turn it upside-down) of a small-mouth 1 quart canning jar or a mayonnaise jar.

You'll also want to make another important purchase from your local hatchery or feed store: a vitamin and electrolyte mix that you mix with the water in tiny amounts. Use it. Unlike antibiotics and hormones, it is not dangerous to you, and it makes growing fast easier on the constitution of your tiny chicks.

At the very beginning, put the water jar within 2 feet of your lightbulb or other heat source so the birds don't have to go too far to find it. Within a couple of weeks you will need to move up to one or more 1-gallon plastic poultry waterers from the same feed store; you'll soon trade this in for one of the metal or plastic waterers that comes in 5- and 10-gallon sizes.

Until your birds reach maturity, keep the water within 10 feet of the chicks' heat source. Replace the water every day and never let it run dry—they can't drink much at once, but they need to drink often in order to grow and prosper. Never use a waterer that the chicks can walk in or sit on top of, because allowing manure to get into the water is the quickest way to spread disease that will kill your young birds.

FEEDING YOUR CHICKS

There are metal canning jar bases for feed, too, as well as trough feeders like the
ones pictured in the brooder setup. In order to discourage the chicks from
standing on top and defecating into the feed, the trough feeders either have a
rolling bar across that top or holes in the top for the chicks to poke their heads
in. The rules for its placement are the same as for the water: no more than 2 feet
from the brooder for the first week or so, and no more than 10 feet away after
that. As the chicks grow, keep the lip of the feeder at about the height of the
chick's back so they will spill and waste less feed. After a few weeks, the chicks
are likely to be heavy and aggressive enough to knock over any of these feeders.
As soon as their necks are long enough to reach over a hanging tube feeder's
high lip, you should hang one an inch or 2 off the ground.

CHICK STARTER AND COCCIDIOSIS

Most commercially produced chick starters include medication to help prevent coccidiosis, the most common ailment afflicting young chickens. It's up to
you whether you want to use this type of feed or find one without medication.
It depends, in part, on the conditions in your coop, the number of birds you
keep, and the climate you're raising chickens in.

Coccidia, the protozoa that cause coccidiosis, live in the gut of every
chicken from the time they are very small. If they multiply at a slow rate inside
the chicks, they don't pose any threat to the birds, which gradually develop
immunity to them in about 14 weeks'
time. Coccidia thrive in damp conditions, such as damp chicken litter, and
are present in large numbers in the
chicks' manure. If the chicks' water is

Opposite: Newly-arrived chicks
from the hatchery get acclimated
to their new home. "Chicks are
low-tech," said one keeper. "They
just need a cardboard pen and a
light bulb on a cord."

Above: One of the newborn hatchery chicks waits for his turn at the feeder in the author's coop.

not kept clean and free of manure, the chicks will swallow additional populations of coccidia, overwhelm their infant immune systems, and get sick. Runny manure and general lack of thrift are some of the earlier symptoms. Bloody manure is one of the later symptoms. If you see these signs and haven't been using the medicated feed, you should remove the affected birds, introduce medicated feed or water, and try to clean the place up. After about 14 weeks, the birds are old enough to have adapted to the local coccidia and are large enough to manage without the drugs.

If you are raising batches of 100 or more chicks, or if you're starting your birds once the weather is already warm, it might be difficult to protect them from coccidia without a coccidiostat in their feed. If the chicks stay warm and dry and their water and litter are kept clean, however, you can raise chicks without medicated starter. It may take some looking to find such a product. Blue

Seal, for instance, sells a 22 percent starter ration called Game Bird Crumbles made without medication that will suffice for young chickens. You may be even better off buying from an organic grain company, although it will likely be more expensive.

Grit

Chicks need to ingest grit, or sand, in order to digest their feed. If they were outside, they would be swallowing sand themselves, but chicken grit is available at the feed store too for a few cents a pound. Sprinkle it on top of their feed the way your kid puts sugar on his cereal and your chicks will get what they need. Grit is discussed in greater detail in Chapter 6.

WHAT TO DO WHEN THE CHICKS ARRIVE

If your chicks are coming by mail, you might want to call ahead and let the post office know you are looking for them. Although at first you may feel odd discussing chickens with your postmaster or -mistress, they have likely had chickens come through the mail before, and they know the drill. If you wait, the post office will call you on the day that they arrive, but I like to be at the post office before it opens instead of waiting for a phone call. It isn't an every-day occurrence, so treat it with the pomp and circumstance that Chick Arrival Day deserves.

Chicks are made by nature to live for 48 hours after hatching without food or water, but it is better that they drink sooner rather than later. After taking them out of the box or out of the incubator, dip the beak of each chick in the mineral-laced water. They will cheep loudly and struggle, but most will drink a few small gulps and seem glad to have had the water pointed out to them. You don't need to treat feed in the same way, but be aware of birds that seem listless and uninterested in the feed. Slow starters can be helped if you dissolve one-third of a cup of sugar in each quart of their water.

After that, just watch and make sure nothing bad happens. Check on them often—certainly no less than twice a day—to replace water and feed. Clean the feeders and waterers. Keep the cat out of the room.

HOW TO DEAL WITH VARIOUS TROUBLES

If just one or two chicks in a batch of twenty-five die before reaching maturity, you are doing well. If they start to drop off with more regularity, you might be looking at a problem. The first thing to watch for is that each chick drinks water, and after that, that they are all eating. If your chicks don't eat or drink, they will die—there is no getting around it. Keep an eye out for a few of the other more common reasons that birds die in addition to coccidiosis: pasting up, picking, and runts.

PASTING UP

Occasionally the trip across the country in the airplane, or simply the stress of being new in the world, can cause the stool of the chick to be looser than it should be. Instead of dropping to the floor to be absorbed in the litter, it sticks to the down immediately surrounding the vent. If allowed to persist, the manure on the bird's bum can harden and seal him up so that he will die of an inability to poop. There's an image that should keep you up at night.

Pasting up can occur as a result of many kinds of stress. If not attributable to travel, pasting may be a sign that the birds are too cold. Pay attention again to their behavior under the brooder, and adjust it accordingly. If the birds show signs of it, clean the paste off with a paper towel moistened with warm water. The bird will complain mightily, but you know it is for his own good.

PICKING

When chicks peck at one another, it is called picking. Picking can happen at any time, but it doesn't have to, and it is easier to avoid it starting than it is to stop it once they get going. The most common reasons for it are overcrowding, overheating, and insufficient air replacement. Chicks may begin to pick the toes or feathers of their mates as a result of these stressors, and once there is blood, it only gets worse. Chicks can literally peck each other to death. If you have adjusted for the three big problems above, try throwing them green grass clippings to distract them from the wounds of their neighbors. Clean the wounds of the victims, and put bacitracin ointment on them to promote healing. A red heat lamp will reduce the chance of picking because it makes it harder to see the toes and other spots that visually attract the pickers. If you have a white lightbulb, put some red cellophane in front of it—being careful that it doesn't melt, stick to the bulb, and burn with a lot of acrid smoke.

RUNTS

Many batches will have a runt or two. These are birds that can't compete and that get shoved aside whenever they approach the feeder. They stay small as the rest of the flock grows. As you raise the feeders farther and farther off the floor in order to keep them at the level of the birds' backs, you may make it harder for the runts to reach the feed. Farming can be a cruel business at times, and this is one of those times. Take the runts out and give them a decent burial instead of watching them slowly starve to death or die of something else because they weren't hardy enough to withstand the local pathogens.

CHAPTER FIVE
Buying Adult Chickens

oultry-keeping is interesting because it gives pleasure as well as profit to persons of both sexes, all ages, all walks of life in all sections of the country. It is useful alike on the farm and in the city back-yard, in the cold North and in the sunny South, in the hill and mountain country, or on the plain. It may be conducted as an exclusive business or as a side line. It affords pleasure and profit for rich and poor alike. In fact, it is the universal agricultural industry. . . . It is a health-giving recreation to thousands as well as a means of support.

—LOUIS HURD, *Modern Poultry Management*, 1928

If you decide to forgo the complications of incubating eggs or brooding chicks, you can still have a fulfilling relationship with chickens by buying adults. You'll want laying birds or cocks if you choose this route—there's little point in buying a meat bird as an adult, as they are ready to eat after only a few weeks. Pullets are sexually mature at 18 to 20 weeks when they start to lay eggs. Cockerels are sexually mature at about 25 weeks.

Opposite: Jim Peavey of East Barnard, Vermont, holds a Silver-Laced Wyandotte rooster he bought for a friend at a sale and swap hosted by the Vermont Bird Fanciers Club.

Above: This hen has found a
new owner at the Vermont Bird
Fanciers Club sale. The group
holds four such events each year.

Where to Look

You'll need to use your own resources to find a grown chicken—I don't know of any mail-order suppliers or warehouses that specialize in adult chickens. Don't despair, however; such birds are surprisingly easy to find. Check with the veterinarian, look in the classified paper, travel to a fair, look in the yellow pages, search the Internet, talk to the butcher at your grocery store, or contact relatives who live a little farther out than you do. You should be able to find someone willing to part with a few hens for $2.00 to $5.00 apiece.

You may even be able to get your birds for free, as it is not unusual for someone with chickens to have some they don't want. A career in poultry is a fluid thing, with periods of ebb and flow, of growth and shrinkage. For all but the most experienced, and even some of them, there are times when it becomes clear that the coop is too crowded, whether because for now it is too much water to carry or the neck, tail, and wing feathers of the birds show too much sign of picking (one of the signals of overcrowding). It is time not only to make soup of the older birds no longer laying, but to find decent homes for the one-and-a-half-year-olds in the middle of their laying years or even the pullets that hatched in the spring and are about to start laying.

There are some risks to finding birds this way, of course. I was told a story once that undoubtedly has been repeated over the millennia in places where rural and urban cultures meet. This particular man got a false start in his life with chickens when a more or less straightforward neighbor offered to sell him four hens for a few dollars apiece, assuring him that the birds would pay for themselves by laying eggs for many years to come. Two of them were evidently older than the buyer realized, however, and died soon after. Although the survivors entertained their keeper for another season or two, only one of them supplied any eggs in return for her room and board. This is not to say that people don't give away perfectly good hens for nothing occasionally if they have small backyard flocks and little interest in recovering all of their external costs. But in choosing chickens, as with any other serious human endeavor, it is well to stay alert.

WHAT TO LOOK FOR

There are a few things to look for in both hens and roosters. Bright eyes; smooth, shiny feathers; smooth, clean shanks (legs); and full, bright, waxy-looking combs. In particular, the legs of an older bird are rough and hold the dirt and manure, whereas the legs of younger birds are smooth, and dirt does not prevent you from telling that the legs are either yellow or white, depending on the breed.

CHECKING FOR PARASITES

If you are buying a hen or pullet and haven't yet learned to grab one yourself, without too much fanfare you can get the seller to hold it upside down. Forget your squeamishness and look at the vent and

Opposite: Michael Whalen of Tunbridge, Vermont, unloads a pickup truck load of chickens to sell and barter at the Vermont Bird Fanciers Club. Whalen raises 25 different breeds of chickens on his dairy farm.

under the wings for external parasites—little bugs crawling on their skin. These are mites, lice, and fleas, and you don't want them in your coop. Keep a lookout also for loose stools that obviously stick to the feathers of the rump, which could indicate worms or other internal parasites.

How To Tell If A Chicken Is Laying Eggs

While the bird is being held upside down, there are a few other things to look for that will tell you whether a hen or pullet is currently laying eggs or not. First look at the vent of the bird again to see if it is moist and oval-shaped, which suggests she might have passed an egg recently. The illustration below compares the vent of a hen that is laying to that of a hen that is not laying.

Next, feel the chicken's underbelly where the fluffy feathers are above the vent. If this area is soft and doughy, it is a good sign. On a bird that is not laying eggs, the skin across that area is tight, or feels thick and coarse. Next look

VENT OF A HEN THAT IS LAYING   VENT OF A HEN THAT IS NOT LAYING

Above: A Rhode Island Red bantam trio, brought by Harry Russell of Hinesburg, Vermont, at the sale and swap.

at the pubic bones, which stick up on either side of the vent. They are wider apart and are soft and pliable on a bird that is laying, and you can fit two or more fingers between them. If you can fit only one finger between the pubic bones, she probably is not laying. In a similar manner, the space in the other direction, up from the pubic bones to the tip of the bird's keel bone, is bigger on a hen or pullet that

WHEN THIS PART OF THE BODY LOSES ITS YELLOW COLOR . . .	. . . THIS MANY EGGS MAY HAVE BEEN LAID
VENT, EYE RING, AND EARLOBE	10
BEAK WITH HALF ITS COLOR	15
BEAK WITH ONE-THIRD ITS COLOR	25
BEAK COMPLETELY BLEACHED	35
BOTTOMS OF FEET	75
FRONT OF SHANKS	95
REAR OF SHANKS	160
TOPS OF TOES	175
BACK OF HOCK JOINT	180

BLEACHING SEQUENCE CHART

is laying. Put your fingers across the abdomen of the bird to measure this width. Two fingers is small and indicates a non-layer; a width of four fingers is what you want to see.

BLEACHING SEQUENCE

Although some breeds of chicken have white skin, most have yellow skin. When a hen begins laying, the minerals that produce the yellow pigment on her legs and in her beak and other areas is diverted to the yolks of the eggs, and the color slowly drains from the pigmented parts of her body. This happens in a prescribed order, both when the pigment disappears during the heavy laying season and when it reappears in reverse order as a hen stops laying during the molt or in the winter. The table above gives a rough idea of how many eggs a bird has laid based on which parts of her body have gone from yellow from white. If you are without experience in looking at the color of the legs of

chickens, it may inspire you in any case to pay attention as your own birds grow older.

FINAL ADVICE

As you look over an adult bird someone wants to sell you, the best advice is not to worry.

Have a critical eye, and use your best judgment. The worst that will happen if your hen is an egg-laying failure or breathes its last on her first night in your coop is that you will lose your $3.00 or $4.00 or $5.00 and have a story to tell.

Above: A New Hampshire pullet gets something to eat from a feeder in Tom Powers' chicken house. Powers said he's had many chickens given to him over the years — "If you have a chicken house, they just show up."

CHAPTER SIX
Feeding Your Chickens

"These animals are omnivorous and there is nothing that they do not devour and consume, assisted by the fierceness of their nature, to such an extent that not only beside almost all kinds of grain they enjoy the flesh of all land and water animals. They do not refrain from even human dung or serpents, scorpions, and poisonous creatures of this kind."

—ULISSE ALDROVANDI, Concerning Domestic Fowl
That Bathe in the Dust—The Chicken,
Male and Female, 1598

What you feed your chickens depends on whether you have meat birds or laying hens, their age, whether they are able to find their own food in the yard, and what commercial feed is available at the local feed store. Layers that are producing an egg every day or so—or baby chicks that can grow from zero to 4 pounds in 8 weeks—need to eat a lot, and they need to eat well.

Opposite: With a piece of hay in her beak, a Dark Brahma hen struts with authority across Carrie Maynard's barnyard. Maynard's birds are free range, which gives them a higher variety of food to choose from.

Above: Frye the piglet eats a morning meal of feed with his feathered friends . The pig and the chickens share household scraps tossed their way, although the birds will peck at the pig if he gets in their way.

What to Feed

Water

Chickens and their eggs each contain about two-thirds water. This water must be replenished constantly, as it is central to many of the processes that keep them alive: from swallowing feed and digesting it, to regulating body temperature, to lubricating joints, to allowing nutrients to move through cell walls. Chicks in particular need water, but at every stage of life the birds need access all the time to clean water—preferably not too warm at the hottest time of the year and not too cold at the coldest.

Commercial Feed

Commercially produced feeds that you can find at your local feed store have been formulated to include all the nutrition your birds need, including vitamins and trace minerals to avoid developing serious health problems. The most common formulas available are a combination chick starter/grower ration in medicated and nonmedicated varieties (the pros and cons of medicated starter are discussed in Chapter 4) and a layer ration.

In rural areas, where the chicken population is larger, more specialized feeds are available. These will allow the keeper to fine-tune the feeding program to the particular stage of development of a bird with a particular purpose. For instance, in some places you can get a "broiler starter," and a "broiler finisher," as well as "grower" and "developer" rations for pullets. The big differences in these formulations is their protein content, because a chicken's protein requirements decrease after the first few weeks. The starter rations contain the most protein (from 21 percent to 24 percent), and the finisher (18–20 percent), grower (16–18 percent), and developer (14–16 percent) rations contain progressively less protein.

Commercial feed comes in three forms: mash, crumbles, and pellets. Mash is simply ground feed. It is not ground to a powder, which is difficult for chickens to eat, but is something like the consistency of sand (although it contains different ingredients of varying sizes). Crumbles are ground feed that has been further processed so that the feed sticks together in small clumps. The clumps are easier for brand-new chicks to pick up and harder for any bird to spread all over the floor. Pellets are the same ground feed compressed into cylindrical pieces that only bigger birds can handle. Crumbles work best for the youngest birds, and although almost any size bird will eat mash, after a chicken is large enough to handle pellets, this is the best choice. The birds will waste less because the pellets don't spill as easily from the feeder.

Differences Among Rations

Chicken feed contains protein, carbohydrate, fat, vitamins, and minerals. The protein is used to make much of the structure of the bird, including bones,

muscles, skin, feathers, and internal organs. Carbohydrates provide energy for fueling everything that happens in the chicken's body. Fat is the form in which that energy is stored for future use. The roles of the various

Above: Tom Powers feeds his chickens coarse cracked corn, which helps the birds produce heat in the colder months.

vitamins and minerals are more complex, and here we will only say that a chicken cannot do without them.

Grains—mostly corn, oats, and wheat—satisfy the carbohydrate requirement and make up the bulk of any commercial poultry feed. The protein portion of chicken feed is usually soybean meal. Although not the largest part of the ration, protein is the most expensive part, and the various rations are usually discussed first in terms of their protein content. When feed people talk about a "16 percent layer ration," it contains 16 percent protein. There are occasionally significant differences in the mineral or other content of the feeds as well. For instance, as discussed later, the layer ration has much more calcium than feeds for other chickens.

KEEPING FEED

Feed does not keep indefinitely. I would suggest that you buy supplies for a couple of weeks or a couple of months—no longer, or you risk it going stale. You'll also want to be on the lookout for rodents, whose appetites will also threaten your store of feed. I used to keep feed in a plastic garbage pail with a lid so it would stay airtight and fresh, until the rats appeared and chewed a hole in the lid. A galvanized metal can with a lid and a bungee cord (or other mechanism to keep the lid on) will both keep the feed fresh and keep unwanted creatures out of it.

SUPPLEMENTING THE COMMERCIAL RATION

Even in summer, particularly if you are a beginner, it is best to keep a commercial ration available to your birds at all times because it guarantees that they have

Above: Chickens owned by Carrie Maynard and Jim Peavey eat their morning meal scattered on and near newly-fallen snow.

access to all the protein, carbo-hydrate, vitamins, and minerals they need. That being said, there are various ways to supplement or adjust their diet in order to cut your feed cost, use feeds that are more easily available than commercial rations, or simply give them some variety to keep them interested and make their meat and eggs more delicious.

Above: A Columbian Wyandotte hen looks for something good to eat in the woods behind the chicken coop on Carrie Maynard's property. Maynard said her chickens will even eat mice and frogs if they cross their path.

FORAGE

If you can let your birds outside, they will supplement for themselves by eating just about anything that is edible. Beetles, buds, bugs, seeds, spilled grain, crickets, worms, termites, grubs, grass, leaves, dropped fruit, flowers, and almost anything they find in the compost pile will serve a chicken. If they are confined and can't get to the compost pile, they will do well if you can provide them with table scraps collected in your

"*U*nconfined chickens find their own food, for the most part, in summer, and do not need to be fed except when they are being fattened for eating, and in the winter months when foraging is meager. The basic point to emphasize is that chickens, like all living creatures, love to be free. They are much happier roaming about than shut up in cages. It does not, of course, follow that they lay more eggs (rather the reverse), but it is certainly true that they enjoy laying them much more, that they lay better eggs, and, for the most part, they are much easier 'keeping.'*"

—CHARLES DANIEL AND PAGE SMITH,
The Chicken Book, 1975

kitchen: vegetable peels and ends, meat scraps, stale bread, the milk at the bottom of the cereal bowl, wilted lettuce, overripe fruit, anything your child leaves half eaten, dog food, cat food. Sour milk is one of their favorites and was regularly fed to chickens in the days when their keepers usually also had milk cows or goats.

A chicken-farming neighbor of mine says that his chickens *are* his composting system. All food waste in the house goes to the chicken coop on a daily basis. The birds don't eat everything (they won't eat citrus fruit, for instance), but what they don't eat gets scratched and trampled until it is part of the litter underfoot that will be next year's garden fertilizer. It is better to withhold anything that is really spoiled, which could make them sick, as well as the onions and garlic, which can pass their flavors on to the eggs and meat.

Your birds will grow more slowly and may lay fewer eggs if they are allowed to forage for themselves, but I would venture to say that it makes for happier chickens. It is also true that judging the relative happiness of a chicken is not as scientific an endeavor as developing a feed ration. Whether it makes the chickens happier or not, you may well be happier watching your hens scratch in the lawn, under the hedge, in the leaves, or wherever they can get to (short of the garden) than you would be if you only see them when you go out to the coop to dump a little more feed in their trough or break the ice out of their water bowl.

GRIT

In chicken terms, grit is small sharp-edged stones or pieces of stone—ranging from the size of grains of sand to the size of small pebbles—that a chicken swallows to aid its digestion. The grit stays in the gizzard, which is a large muscular pouch just above the intestine that grinds the grit together with the chicken's food. "Scarce as hen's teeth" is how the saying goes, because chickens don't have teeth at all. They rely instead on the gizzard and grit. Chickens that only eat commercially prepared rations do not absolutely need grit because the feed has already been ground. Chickens that are fed any whole or cracked grains have to have grit available to them to do the grinding. When left to their own devices outside, chickens swallow small stones on their own. If kept in confinement, or in colder climates where the ground is out of reach under the snow in winter, you need

 Opposite: One of Tom Powers' hens wanders outside the open gate to her pen. Powers lets his chickens roam around the yard when the garden isn't ready for harvest.

Above: A hen takes a cruise around the open land. Chickens supplement their feed with insects and small reptiles found while free ranging.

to supply your birds with whatever grit is carried by your feed store. Clean gravel from a local streambed works just as well.

USING OTHER AVAILABLE FEEDS

With some experience and practiced observation, you can eventually begin to adjust the protein level in the ration yourself in order to cut costs and use grains and other animal feeds that are available to you. I have a friend with a cow who buys a growing mash with the lowest protein—15 percent—for his chicks but feeds them milk instead of water to add what for him is free protein back into the chicks' diet. They grow a little bit more slowly than they do on chick starter alone, but it's worth it to him: He gets to use what he has on hand.

PEARSON'S SQUARE

Pearson's square is a method you can use to make simple adjustments to the protein content of your main ration. In the example below, for instance, the aim is to figure out how much commercial chick starter/grower containing 22 percent protein should be combined with how many oats, which are 12 percent protein, if you want to end up with a "developer" ration containing 15 percent protein for your young pullets before they begin to lay eggs.

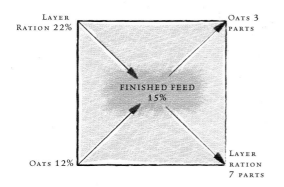

PEARSON'S SQUARE

The protein content you are aiming for is in the center of the square. On the left are the protein contents of the two feeds you will mix, chick starter on the top and oats on the bottom. You can find the protein content of the chick starter printed on the label attached to the feed bag. The protein content of oats and other feeds can be had from the closest extension office (Colorado State University has a particularly good bulletin on home feeding of poultry at www.ext.colostate.edu/pubs/livestk/02504.pdf).

Once the numbers are in place, you can forget the percentage signs and simply calculate the differences between each of the numbers on the left with the number in the middle. On the right side of the square you are left with the proportion in which to mix your two feeds. In this case, the difference between 15 and 22 is 7, so you need 7 parts chick starter. The difference between 15 and 12 is 3, so you need 3 parts oats.

FORMULATING YOUR OWN FEED

Raising your own feed is an option, but not a very good one for the beginner. Growing feed is a considerable effort and expense unless you already have the time, inclination, and equipment to do it. In any case, mixing feed with the correct protein, carbohydrate, vitamin, and mineral content is more easily done in the laboratory. Still, there are social, political, and economic reasons for creating your own chicken feed that deserve serious attention, from the use of pesticides, petrochemicals, and patented seed in the production of the feed to the inclusion of medications in the rations. For more information on creating your own rations, see www.lionsgrip.com/chickens.html.

THE PRACTICE OF FEEDING

Although some prefer to bring individual meals to their chickens once or twice a day, free-choice feeding is the preferred method of backyard chicken raisers. This means that the birds have food in their feeder at all times and take it when they want it. The birds will waste more feed than they should, however, if you don't let them eat all the food in the feeder before adding more, at least once in a while. Chicks in particular should have food available to them at all times, because they are growing so fast.

Above: Tom Power's rooster heads for the cover of overgrown brush in the barnyard near the henhouse. The rooster, like most of the birds Powers has, was given to him.

FEEDING MEAT BIRDS

The common rule of thumb says that you will feed 2 pounds of feed for every pound of weight that your broilers gain. If you want to grow twenty-five birds to 4 pounds each, you should plan on buying about 200 pounds of feed before it's all over, at a total cost in the neighborhood of $30.00.

Start chicks off with a commercial chick starter/grower feed. When a chick is only a few days old it will take him what seems like forever to go through a pound of feed. Your first 50-pound sack will last for the first couple of weeks. Rest assured, their appetites grow as fast as they do. If it is warm enough and you can give them access to the outside, open the door for them when they are about 2 weeks old during the warmest part of the day. Meat birds are not as interested as other chickens are in eating grass and bugs, but they will relish them if they learn about them early.

If a finishing ration is available at your local feed store, begin to give it to your chicks when they are 3 or 4 weeks old by mixing a little more each day with the starter that they are used to until you have completely switched to the finisher over the course of a week. A finishing ration contains more carbohydrate than the starter ration. This allows the bird to put on a little more fat and, as the bird's protein requirements begin to decrease, makes it slightly cheaper to feed the birds for the last 2 to 4 weeks. If finisher is not available, you can slowly mix in cracked corn or other "scratch" feed, which is easier to find than finisher. This has the similar effect of cutting the protein content and increasing the carbohydrate content of the feed. Add the corn until it is one-quarter or one-third of the total feed ration.

FEEDING LAYERS

Once they are full grown and laying eggs, the amount an adult laying hen eats varies with its age and size, the air temperature, and the number of eggs it is laying, but a rough average is a quarter pound of feed per chicken per day. Under this rule, a dozen layers will eat two 50-pound sacks of layer pellets in a month, which might cost you $12.00 to $15.00.

Opposite: A Dark Brahma hen gets a drink of water in Carrie Maynard's barnyard. Maynard said her 38 birds drink about three gallons of water a day, which keeps them hydrated and their skin moist.

Above: Max the Light Brahma rooster explores a field of chopped corn next to owner Lianne Thomashow's home. Thomashow's dozen chickens wander around the yard for most of the day.

If you can let them outside, treat the layers the same as the meat birds, encouraging them to forage after about 2 weeks, all the while feeding chick starter, or pullet starter if it is available. If you can get pullet grower, gradually switch to it after 6 weeks. If pullet developer is to be had, gradually switch to that after about 12 weeks. Finally, switch over to a layer ration after 20 weeks.

If that sounds complicated and all you can find is chick starter ration and layer ration, don't worry. You can provide perfectly acceptable nutrition for your pullets by mixing oats—which are much more commonly available than the various pullet rations—into the chick starter. This is the example we used to illustrate the workings of Pearson's square above. Don't add the oats until the birds are 8 weeks old. At that point, begin by mixing one part oats to ten parts starter ration, and increase the proportion one part at a time every day until the feed is about one-third oats. When the birds are 18 weeks old, make the same gradual switch to a layer ration.

CALCIUM

Eggshells are made from calcium, which comes from the hen's diet. Although their commercial ration has plenty of calcium for an average hen on an average day, calcium requirements for any particular laying hen go up and down depending on their age, the weather, and the season. You will therefore want to provide a source of additional calcium for your laying flock. I suggest using pulverized oyster shell, available in most feed stores, for this purpose. If you prefer not to buy the small feeder that can dispense the ground shells, put some in a small bowl or in a pile on one side of the feeder for the birds to take if they want it. Crushed eggshells will do as well if they are clean and broken beyond recognition. Feeding eggshells that look and taste too much like eggshells might encourage the hens to eat their own eggs.

FEEDING BOTH LAYERS AND MEAT BIRDS

Here is a typical scenario for the backyard operation starting twenty-five chicks that will be half layer hens and half meat birds. Feed them all chick starter together. Introduce some cracked corn or oats after 5 or 6 weeks, and kill the meat birds when they are 8 to 10 weeks old. Over the next couple of weeks, add oats to the pullets' feed until it is one-third oats. When they are 18 or 20 weeks old, switch them slowly to a layer ration.

Grass Feeding

Although many people let their chickens run in the yard or in a fenced-in chicken run, you can use pasture more efficiently and even improve it by keeping your chickens in a confined space on the grass and moving them every day or every few days. You will still need to feed a balanced feed ration, but your birds will eat less of it if the grass is young and tender, the way they like it. As you move them from one patch to the next on a regular basis, their manure will be distributed evenly over the ground in amounts that will feed the grass instead of burning it. The scratching that the chickens do will be enough to aerate the soil near the top, but not enough to dig the plants up and kill them off, as happens over time in the run.

There are two basic options for grazing your chickens. The first, less intensive method is a coop on wheels. It is easily moved from one spot to the next inside a paddock made with portable electric fencing to keep the birds enclosed in a large area surrounding the coop. The portable fencing is lightweight electroplastic strands or webbing. The wire is hung on fiberglass posts that are easily pulled up or pushed into the ground with your hands (see the Appendix for sources of fencing). The coop provides shade by day and safety from predators at night. If you have layers, they can have their nesting boxes in there, too. Move the fence and coop every few days when the paddock begins to look weary and bare spots begin to show.

The second option is a low flat cage a couple of feet high that keeps the birds confined in a limited area of perhaps 2 square feet per bird. This method is generally used for meat birds. In the space of 24 hours, they eat the grass down close to the ground, and they supplement their protein intake with insects they find in the grass. You must move this assembly each day, but this is the way to get the most out of the pasture or lawn. Grass as a species evolved in large flat

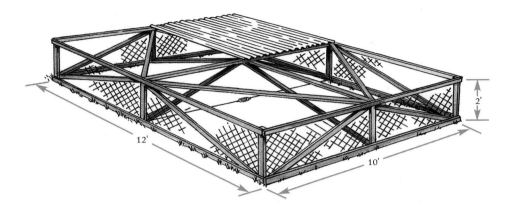

Moveable pen for intensive poultry grazing pioneered by Joel Salatin.

expanses where herding types of animals traveled in closely bunched groups for safety. Crowded together, they ate all the plant material down in a small area and then moved on to another area, leaving the eaten plants to grow back. By "mob-stocking" one small patch at a time and moving your chickens daily, the grass gets the competitive advantage over the weeds, and the pasture only gets better and better.

FEEDING EQUIPMENT

Chickens are well equipped to feed themselves, but chicken farmers like to have a place to put the feed so it is not spilled and lost. Feed represents the bulk of the cost of keeping the birds: about 60 to 70 percent. Most of us rely on simple troughs, either on the ground or hanging from the ceiling by chains. Also important is getting water to the chickens: Unless you can provide a lake, creek, or rivulet by the coop, you'll want to have some sort of container for this.

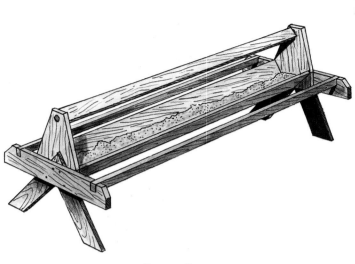

TROUGH FEEDER

TROUGH FEEDERS

The trough feeder is the most common type, and there are various styles available. Each is designed to prevent the chickens from stepping in or roosting over the feed and soiling it with their manure. Each is made to adjust higher as the birds grow, some suspended with chains, others with adjustable legs. Some have roosts built into them to allow the birds to sit and eat. Don't fill them much more than half full, or the birds will spill as much as they eat.

TUBE FEEDERS

As far as I'm concerned, none of the trough feeders are as useful as the hanging tube-style feeders. First, tube feeders are easy to adjust in height as the birds grow, as they are suspended by a single chain and an S-hook. Because they operate on the first-in, first-out inventory method—which in this case means that the chickens feed from the bottom—new feed is added from the top and will only be eaten after all, or most, of the older feed on the bottom is gone.

TUBE FEEDER

Both trough and tube feeders can be used with chicks as soon as the chick is as tall as the lip of the feeder. Keep the top of the lip roughly at the height of the birds' backs. The chickens will spill less if they are not reaching down low to scoop up each mouthful of feed.

WATERING METHODS AND WATERERS

Watering methods vary according to the patience, strength, and ingenuity of the keeper as well as the climate that keeper lives in. Carrying water is not difficult if your flock is small and your source is not far. Keep in mind that you need to scrub the waterer anywhere from twice daily to once weekly, depending on how many birds are dirtying it and how hot the weather is. There are a number of different types of water containers, or *waterers*, to choose from.

Conventional chicken waterers available at the feed store are made of metal or plastic and come in quart, gallon, and 5- and 10-gallon sizes. The metal ones

WATERER

will probably last longer, as the plastic ones tend to get brittle and develop cracks or shatter altogether if you don't treat them with care. Choose your size and number of waterers in order to allow at least one-third of your birds to drink at once, no matter what kind you have. If the area around the waterer tends to get soggy, you can build a wooden frame of 2-by-4s with heavy wire mesh stapled to the top of it; the chickens can hop up onto the mesh to drink. Like the feeders, keep the lip of the waterers at a height level with a chicken's back.

If you don't want to deal with delivering water twice a day, automatic waterers that are constantly fed from a pipe or hose are available in feed stores and on the Internet. If you live in a cold climate, bury a pipe below the frost line and have it enter the coop in a way that will prevent it from freezing on the way in. This will save much effort. One type of automatic waterer is a cup with a valve in it that lets more water in when the level in the cup gets too low. Another is a nipple waterer, which amounts to a chicken water fountain that the birds sip from directly. This is a luxury item—automatic waterers will save the chickens the discomfort of a waterer gone dry on a hot day and prevent them from going thirsty on mornings when their water is frozen solid and you are sleeping in. In addition to all of this, it will save you from having to ask a neighbor to carry water and break ice out of the waterer if you go out of town.

If you live where the water will freeze, there are various approaches to watering. The time-honored method in the winter is to go to the coop twice a day with a new container of fresh water and swap it out with the frozen one, which you bring inside to thaw out in a utility sink somewhere, or on the floor in a corner of the basement. Before I replaced rubber and plastic bowls with a regular poultry waterer, I had a certain amount of success carrying two gallon plastic milk jugs full of hot tap water to the coop. I poured one gallon on the back of the bowl, warming it enough to pop the ice out, and the other gallon to refill.

One of my neighbors, after years of breaking out ice, decided that his bantams could get all their water in the winter from eating snow. A more conventional method is to keep an electrical warming device (made expressly for the purpose and available at the feed store) under the chickens' waterer. I have also seen a similar device made at home from a ceramic light socket attached to the inside of the lid of an empty glass peanut butter jar such that a 40-watt lightbulb was lit inside the jar. The jar was put in the space inside a cinder block laid on its side. The waterer was set on the cinder block, and the lightbulb underneath kept the water from freezing.

CHAPTER SEVEN
Hens and Eggs

"*The question (of which came first, the chicken or the egg) simply falls away when one considers the chicken and the egg as processes as well as objects. Each represents not a separate unit, but transitional stages in a cycle of life, all parts of which are molded by evolutionary pressures.... The egg, representing the female component of the germline, is the bridge between individual cycles and is present in some form at all stages. Thus even when the chick is in the egg there are eggs within the chick, microscopically small but full of potential.*"

—CHARLES DANIEL AND PAGE SMITH,
The Chicken Book, 1975

Young pullets begin laying at 18 to 20 weeks of age and after a somewhat slow start should lay continually until they molt (lose their feathers) for the first time about a year later. No chicken will lay as many eggs in her second year as in her first, and although some may continue to lay slowly and steadily for 5 to 10 years or more, most will probably eat more than they are worth in eggs after the first couple of years. After that they go to the stew pot. The stock made from stewing old hens past laying is

Opposite: This hen doesn't seem to mind the fresh snow on a cold afternoon.

the chicken soup known across cultures and generations to cure colds and other ailments.

The Formation of the Egg

A hen's willingness and ability to lay on a given day is subject to her physical condition, what she has been eating, and the season (light plays a critical role). As explained in Chapter 3, a typical cycle calls for a hen to lay an egg about an hour later each day, until the laying time starts to run into the evening hours. The hen won't lay in the dark, so at this point she will rest overnight. Unless you are letting her brood, she begins the cycle anew the next morning. This results in something less than two dozen eggs a month—a figure that goes down in the "darker" months of fall and winter.

Every female chicken is born with two ovaries, but because of the room an egg requires, the right ovary stops developing when the hen is young. The left ovary, made up of a cluster of follicles, contains the primordial germ cells that produce eggs.

These cells appear as a cluster of spheres each independently attached to a slender stalk. Imagine them like grapes or berries on a stalk, most very small, some growing slowly, and perhaps a half dozen of different sizes all ballooning at great clip before they pop out of their follicle casing and head down the oviduct in what is called ovulation. That ballooning happens as the yolk material is accumulated in layers, almost all of it in the week before ovulation.

The oviduct is a long passageway coiled in the body of the hen that includes several distinct sections, each with its own role in the formation of the egg. The infundibulum is the first section, where fertilization happens if fertilization is to occur. The ovum is in the infundibulum for only about 15 minutes before it moves on to the magnum, where the layers of albumin—or egg

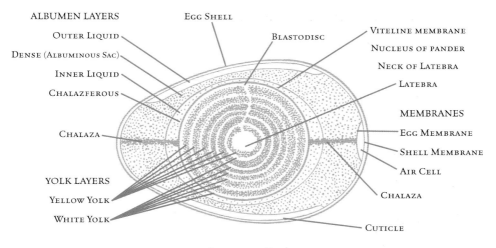

ALBUMEN LAYERS
OUTER LIQUID
DENSE (ALBUMINOUS SAC)
INNER LIQUID
CHALAZFEROUS

CHALAZA

YOLK LAYERS
YELLOW YOLK
WHITE YOLK

EGG SHELL
BLASTODISC

VITELINE MEMBRANE
NUCLEUS OF PANDER
NECK OF LATEBRA
LATEBRA

MEMBRANES
EGG MEMBRANE
SHELL MEMBRANE
AIR CELL

CHALAZA

CUTICLE

PARTS OF AN EGG

white—are wrapped around it. The yolk twists through the magnum, and the thick part of the albumin is twisted at each end like the wrapper on a piece of candy. This process takes about 3 hours. The next section is called the isthmus, where the shell membranes (two of them) are added. These are the tough rubbery membranes that keep the shell stuck together when you peel a boiled egg or crack a fresh one into the frying pan. When the airspace forms at the large end of the egg, it will form between these two layers. It takes a little more than an hour for these membranes to be added and for the egg to move on to the uterus, or shell gland. As the name implies, this is where the shell is applied around the shell membranes over the course of the next 20 hours or so. There the egg waits, once completed, until the hormone secretion and muscle contractions commence that push it out into the nest.

The mucuslike lubricant that allows the egg to exit the hen also helps protect the egg when it dries on the shell and forms what is called the bloom. You can't see the bloom, but it is another deterrent to bacteria that would enter through the pores in the shell, and it helps keep moisture in.

As a hen gets older, the shells on her eggs can be wrinkled. There is no reason not to eat them—they just look funny.

Sometimes you will find eggs with bloody shells. This most often happens in pullets laying their first few eggs before they are quite ready. In most cases, it is not a problem. It is rare that a grown pullet or hen will get coccidiosis, but bloody eggs may be a symptom if bloody stools are found as well.

Eggs are occasionally laid with no shell. You will find them in a puddle, surprisingly complete with yolk, white, and shell membrane, just no shell. This is nothing to worry about if it doesn't happen often.

THE ROOSTER

You don't need to have a rooster in order to have eggs. This is occasionally a matter of confusion to those unfamiliar with the ways of chickens (and the world). You need a rooster if you hope to have chicks hatch from those eggs. There are those who prefer to eat fertile eggs, feeling that these eggs are more complete in some psycho-spiritual-nutritional way, and I will not argue with that. Others believe an egg with a living embryo in it, barely if at all visible to the human eye, detracts from its appeal as food. This is a matter of taste—not of the tongue—but of the sensibility.

Opposite: Roman the bantam rooster crows at Lianne Thomashow's home. The rooster and three of his hens were given to Thomashow by a neighbor. She built a coop out of old doors and window shutters she found in the family's barn.

CONTROLLING LIGHT

Hens need light to produce eggs. Light stimulates the pituitary gland, which stimulates the ovaries and makes for maximum egg production. In the height of summer, nature helps out the chicken farmer with sunlight, but as the number of hours of daylight

Above: A rooster takes a
stroll in the barnyard of
Tom Powers' chicken
house.

drops to 14 in a day—as it does in most parts of the United States from September or October through April or May—a hen may stop laying altogether. Commercial egg producers often use artificial light to support egg production and profitability through the cold, unproductive winter months. Although it is not necessary, and although some hens will continue to lay all winter anyway, you too can increase your egg production by providing artificial light.

For a backyard operation, a 40-watt bulb suspended about 7 feet off the floor will provide light of enough intensity to substitute for daylight in a small coop with about 100 square feet of floor space. If your coop is a little bigger, a 60-watt bulb will suffice for 200 square feet.

Although you have the choice of increasing the light in the morning or in the evening (or some combination), it is better to have the lights come on before sunrise in order to extend the chicken's productive day at the beginning rather than at the end. If the lights go out all of a sudden in the evening while they are busy eating or pecking, they may panic, and they will have difficulty finding their roost in the dark.

You can turn the lights on and off yourself, but you must do it at the right time at both ends of the day, every day. Irregular lighting can cause the hens to go into a molt (more on this later in the chapter) and stop laying altogether. It will probably be easier to install a timer that will turn the lights on and off for you. Keep in mind that you will still need to adjust the time that the lights come on as darkness comes progressively sooner, until the winter solstice on December 21 and progressively later after.

STORING EGGS

If eggs are collected within a few hours of the time they are laid, and if they are kept in a cool moist place at about 75 percent humidity, they can last for up to 4 months. Your refrigerator is much drier than this, which is the reason eggs will only keep for 4 or 5 weeks there, unless you wrap them in plastic to keep them from losing moisture. Keeping the eggs clean is the main reason to collect them often. Washing eggs reduces their shelf life to perhaps a month because it removes the bloom, which keeps them from drying out.

When you are awash in eggs in the summer, the best way to deal with the surplus is to make friends and a little bit of money by selling the excess to neighbors and coworkers. You can also freeze them to use later. To do this, break the eggs into freezable containers and mix lightly; add a dash of sugar or salt to help maintain the texture when they thaw. You can also separate the whites from the yolks; sugar or salt in the yolks will keep the texture from getting gummy.

THE MOLT

The feathers on a chicken, like those of other birds, wear out over the course of the year and are replaced annually in what is called the molt. You may notice your birds looking pretty ragged at some point in the fall, and although you should make sure they are not being pecked or losing their feathers for other reasons (see Chapter 9), this is perfectly normal. They are shedding their feathers and growing new ones.

Above: A Silver-Laced Wyandotte hen is molting at Carrie Maynsard's barnyard. Laying birds usually lose their feathers once a year, but Maynard's birds lost them twice due to unseasonably warm fall weather. The hens stop laying while they are molting—a process that can take weeks to months.

The molt is of particular importance to laying hens. It coincides with the time of year that the hens stop laying eggs and rest for a period that lasts as

little as a few weeks and as much as several months.

Commercial egg-laying operations often replace the hens at this point, having kept them for just one laying year, because they can't afford to feed them without getting eggs in return. Sometimes they "force" the molt by limiting the amount of light, feed, and water available to the hens. If the molt can be completed earlier, laying through the winter can be extended. Like adding artificial light, the forced molt is a way for larger operations to be profitable in the winter, when hens managed less intensively will lay many fewer eggs than they do in the summer.

The timing and duration of the molt also says something about how productive the hen is. Birds that begin to molt earlier without being forced tend to have longer periods when they are not laying and therefore lay fewer eggs over the course of the year. The poorest layers may begin molting in July or August and stop laying for 4 or 5 months. These are the birds to stew when the time comes for stewing. The best layers tend to begin the molt later and molt more quickly. Some will even continue to lay while the molt is in progress and may only stop laying for a period of a few weeks in the late fall.

A molting chicken sheds its old feathers and acquires its new ones in a prescribed order: head, neck, body (including the breast, back, and abdomen), wing, and finally the tail. The ten primary flight feathers that every chicken has at the end of each wing are particularly predictable and allow the careful keeper to gauge the length of time of the molt and predict how quickly it might be finished. A slow molter loses just one primary at a time. Each of these feathers takes 6 weeks to grow back, and they are usually shed at 2-week intervals. A quick-molting bird will shed the feathers in groups of two or three every 2 weeks instead of one at a time, and so reduce the total amount of time needed for the molt. If you see new feathers of different lengths coming in on your bird's wing, she is a slower molter. If the new feathers come in such that two or three are the same length, she is a faster molter.

CHAPTER EIGHT
Butchering Your Meat Birds

" *ho does not know that the chicken is used for*
food? This creature almost alone is our chief
resource when friends or guests arrive suddenly
and unexpectedly; we owe to the chicken all the splendor displayed
by a rich table or by one that is modestly supplied or by that
which is slenderly laden. "

—ULISSE ALDROVANDI, Concerning Domestic Fowl
That Bathe in the Dust—The Chicken,
Male and Female, 1598

There is dirt and physical labor associated with some aspects of raising chickens. For the backyard smallholder, the relative unpleasantness is very limited and should not intimidate anyone with an inclination toward keeping fowl. For those who don't already enjoy such tasks as carrying feed and water or shoveling manure, the company of the birds and the eggs and meat they provide will more than outweigh most of what might be the less attractive requirements of chicken husbandry.

Killing the birds, however, may be another matter. For most of us, it has been at least a generation, or several, since the act of slaughter was a

Opposite: A rooster stretches and shows his might during a warm afternoon at Tom Powers' home. Powers thinks having more than one rooster per flock is "worse than a Western town with outlaws."

Above: Carrie Maynard's Silver-Laced Wyandotte hen scurries across the barnyard. Maynard has had chickens for the past three years and is now an officer in the Vermont Bird Fanciers Club.

regular practice in our families, and both the knowledge of the mechanics and the willingness to do it have been lost. This is reason enough to try it yourself.

The hardest part for the beginner is taking the knife to the neck of a living bird while it blinks at you as it hangs upside down from a string wrapped around its feet; and after, when the blood

drains, and the bird jerks and flaps in reflex. These are confusing, terrifying moments, as well they should be.

BUTCHERING AT HOME

You can process small numbers of birds at home without anything fancier than a sharp boning knife. A "small number" is anywhere from 5 to 150, depending on how much practice you have had and how much help you can find. If you have never butchered birds before, start with five or less.

The basic steps are to kill and bleed the bird, scald it in hot water for a minute or two, pick all the feathers off, pull the guts out, and then put the carcass in cold water to cool.

PREPARATION

The first step is to withdraw feed, but not water, from your birds for 12 hours before slaughter— typically, that means the night before you butcher. The crop and bowels will empty out, and there will be less manure that you have to direct away from the meat during processing.

The next morning, you need to create a place to work. Cleanup is easier if you are butchering outside because spattered blood and bits of skin and feathers will dry up and blow away or be eaten by resident small animals. You will need a table of some kind at waist height that can be cleaned well. Either clean boards or plywood on sawhorses works fine.

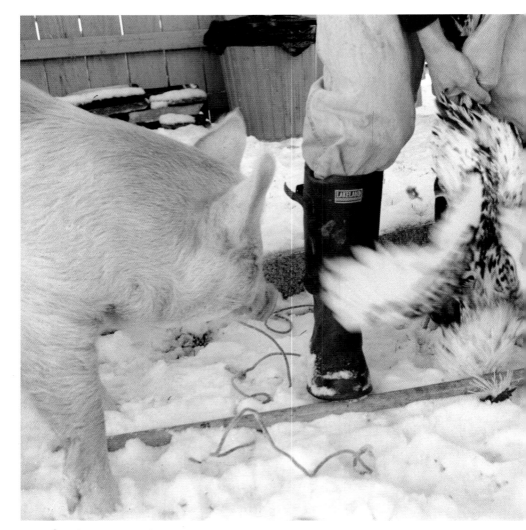

Above: While Frye the pig watches, Lianne Thomashow kills a rooster for butchering by holding its neck down with a broom handle. Thomashow then bleeds out, plucks, and dresses the bird while hanging it from a hook in the fence at her home.

You'll also need to assemble some equipment. A very sharp knife is the most important tool—a dull blade multiplies many times the effort required for processing. For scalding, you'll also want a pot or pail that holds at least 3 gallons and a thermometer to measure the water temperature. It can take a little while to heat 3 gallons of water, and you will want to heat some extra just in case, so put the water on the stove early. Finally, you will need another large vessel filled with ice water to cool the birds after they are dressed. This one should be big enough to hold all of your finished birds at once. Make sure you have plenty of ice on hand, as you will want to allow several hours for the birds to cool.

Killing

Your main object when killing the bird, besides getting it done quickly, is to do it in a way that allows the carcass to bleed out completely. A poorly bled chicken is less appetizing and more subject to spoilage than one that is completely bled out. It is easiest to bleed the bird from the neck, upside down. You must hang the bird by its feet from a tree branch or clothesline, using a slipknot at the end of a short length of twine heavy enough to hold the bird. Better yet is a killing cone, which is a kind of metal funnel big enough to hold the chicken's body that

LETTING SOMEONE
ELSE DO IT

IF YOU DON'T THINK YOU CAN HANDLE killing your
own birds—or don't want to spend the time learning how—
you can often find someone else to help you out, at a cost of
$3.00 to $4.00 a bird. In some places, a local butcher will kill
chickens, or there may be a chicken farmer in the area who
dresses neighbors' birds as well as his own. In other areas,
there are commercial poultry processors who will do your
birds while they are doing theirs. In my area, there is a man
who will bring his poultry-processing apparatus, mounted in
a trailer, to your door and perform the entire operation on
site. You find out about these kinds of people by talking to
others like you who have a few chickens.

    If there is someone you can take them to, be sure to
transport your chickens in wooden crates or cages, five or ten
to a crate. If you close them up in a cardboard box without
putting a lot of holes in it, or crowd too many into too small
a space, they may well suffocate. In large groups in the back
of a truck they can panic, pile on top of each other in a corner, and asphyxiate the ones on the bottom of the pile.

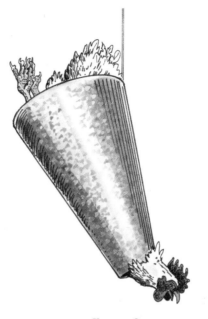

KILLING CONE

allows just its head and neck to stick out of the small end. The killing cone helps prevent the bruising that can occur when the bird flaps and struggles while it bleeds out.

Most methods of killing chickens, including wringing the neck or chopping the head off with an axe, sever the spinal cord and stop the heart immediately. It is more effective to cut the jugular vein and allow the heart to pump the blood out of the chicken. To do this, hold the chicken's head with one hand, and with a swift firm stroke of the knife, cut into the neck immediately behind the jaw, without cutting into the windpipe, esophagus, or spine. The bird will bleed to death quickly, and then flap and twist for another 2 or 3 minutes. If you are using string instead of a cone and can hold the bird's head until the contractions stop, it will keep blood from being spattered all over. Bleed the birds into a bucket for composting.

SCALDING

The reason for scalding the chicken is to make it easier to pluck the feathers. You will need a vessel large enough to dunk a bird in without spilling the water out, and you will need a thermometer. Recommended temperatures for the scald range from 125 to 150 degrees F, but in practice, maintaining the correct temperature is a delicate and critical balance. If the water is too hot, you will find the skin ripping as you pull feathers out. If it is not warm enough, the feathers will be more difficult to pull, and the amount of time and effort needed to pluck the bird—particularly the tail and large wing feathers—will increase dramatically. A good rule of thumb is to scald at 140 degrees for 1 minute. Holding onto the bird by its shanks, dunk it upside down in the water, and whoosh it up and down a bit to make sure it gets soaked through to the skin. Pull the bird out and see how easily a tail feather comes out. With a little practice, you will know whether to put the bird back in the scald for a few more seconds or not.

Depending on how cold it is in the room or shed or yard where you are working, the water can cool enough to do you no good after just a few birds are dunked. You could purchase or build a vat heated with electricity or gas, complete with a thermostat to keep the water always at the temperature you want, but unless you are doing hundreds of birds, this is too expensive to be practical. I keep an extra pot of hot water on a camp stove to heat up the scalding vat. You can also keep extra water on the kitchen range to carry out if necessary.

PLUCKING

Particularly for the beginning home poultryperson, plucking is the most time-consuming and painstaking part of the process. Most of the feathers should come easily in handfuls in a few minutes,

Opposite: Lianne Thomashow plucks the feathers from a rooster she butchered at her home. Thomashow had to kill the bird because she had too many roosters, and the bird was causing trouble at the town clerk's office.

CUTTING OFF THE FEET

but the final cleaning of the pinfeathers, which are the tips of new feathers, can take a few minutes more. Tweezers or needle-nose pliers can help, or you can pull them out between your thumb and a small knife. Stainless steel plucking machines with rubber fingers can pick four chickens clean in 15 seconds, but they are not economical for the small home producer—unless you can convince several of your neighbors to raise their own chickens and go in with you on the equipment.

EVISCERATING

The first step in preparing to eviscerate the bird is to take off the feet and head. Cut the feet off at the hock, which is the first joint. Your aim is not to cut through bone, but to separate the joint. There is a tendon that holds it together that you can cut through by slicing around the back of the joint from one side

CUTTING OUT OIL GLAND

to the other. The hock will separate, and you can cut or pull it apart easily. Similarly, if you cut around the neck, you should be able to twist the head off without having to cut through the spine with your knife.

Now slit the skin along the back of the neck lengthwise, and separate the windpipe and esophagus from the neck skin. If you withheld feed, the crop—which is the enlarged part of the esophagus at the base of the neck—should be empty and small. Cut it off at the base of the neck.

Next, take the oil gland out of the back of the tail. You can locate it by its small oil gland nipple. Cut across the tail about an inch below the nipple toward the body of the bird, and slice down to the vertebra in the tail and then back up to scoop out the gland.

Now turn the chicken over so that it is lying on its back again, with its drumsticks sticking up the way it does in the roasting pan. Make a cut through

BELLY INCISION

the skin and fat layers in the soft belly of the bird above the vent. The cut should be an inch or 2 in length; it should be vertical if you are going to cut the bird into parts, or horizontal if you will roast it as it is. You are being careful at this point not to cut deeply enough to nick or cut open the intestines coiled in the belly. Once you have an opening into the body cavity and can see those intestines, you can tear or cut the opening further until you can fit your hand inside the bird.

Put your hand inside and reach all the way up inside along the keel—which right now is the top of the body cavity—to the top where the esophagus and windpipe come through from the neck. Hook them with your finger and simultaneously grab the gizzard, which is the only really solid organ in the body cavity (bigger than a golf ball and smaller than a baseball). All of the innards should pull out in a mass. At this point, it is all still attached by the large intestine to the

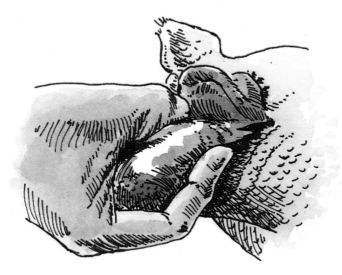

Pulling Out the Guts

vent, so be careful not to pull the intestine apart and spill manure in your
chicken. Rather, cut all the way around the vent so that it comes free from the
bird with the intestine.

Next locate the unmistakably bright green gallbladder, which is the color
of nothing else inside the chicken and shows up obviously against the maroon
liver. Pinch or cut it off the liver, being even more careful with it than with the
intestine, as the bile in it will ruin the meat if it spills. Now you can set the liver
and the heart in ice water and pull off the gizzard if you want to keep it for
gravy or stock. Clean it by cutting it open and peeling the thick, yellow, rubbery
lining off. Throw the lining away with whatever gravel and undigested grain is
still inside it.

Finally, you must reach into the bird a second time in order to scrape the
lungs out. They are pink and spongy and are attached to back of the bird's ribs,

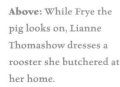

Above: While Frye the pig looks on, Lianne Thomashow dresses a rooster she butchered at her home.

which are now on the bottom of the body cavity, as the carcass is still lying on its back.

Now rinse the inside and submerge your chicken in ice water. Your goal is to lower the temperature of the carcass to less than 40 degrees F as quickly as you can. You should allow 6 to 10 hours for this—less for smaller birds and more for larger.

FREEZING

Your fryers and roasters will be better preserved if they age in their ice water or in the refrigerator for even a few hours more than it takes to cool them—and for as long as 3 days—before freezing. It is also best that the birds be as dry as possible when they freeze in order to avoid freezer burn. Take the chickens out of the ice water, and let them drain for about 20 minutes before bagging them. If you double-bag them in plastic, they should keep for as long as a year in the freezer. If you cut them into parts, don't plan to keep them longer than about 6 months.

CHAPTER NINE
The Health of Your Chickens

"'*M*any diseases can be prevented through good
management but cannot be cured once
they occur.*"*

—University of Missouri
Agriculture Guide G8350, 1997

The health of your chickens will depend
mostly on your management of their keeping
and feeding as discussed throughout this book, but
partly on luck. If you provide them with plenty of space, good light, lots of
fresh air, protection from drafts, access to the outside, proper feed, and clean
water, your troubles will be minimal.

That said, some of your chickens are bound to get sick, and some will die,
often for reasons that remain a mystery. Even a well-run operation loses from 5
to 10 percent of its birds each year. The time to worry comes when three or four
die in a period of hours or days.

You'll need to watch out for disease (invisible virus and bacteria that invade
your chickens' internal systems), parasites
(small but visible creatures that prey on your
chickens), and cannibalism (chickens that prey
on each other).

Opposite: Roman the
bantam rooster stays nice and
dry inside the coop during a
stormy afternoon.

No matter what its ailment, a sick bird should be quickly isolated from the rest of the flock, either in a separate part of the coop or in a separate hospital coop. This will help the bird to recover (if it is going to recover) and might help keep the others from being exposed to whatever is ailing the one.

Rodents have been mentioned elsewhere in this book, but they deserve particular mention here as they can carry diseases and external parasites. The most important protection is to keep your coop and its surrounding areas tidy. Rats are attracted to waste feed and other garbage, and they gain easy access if they can hide in tall grass or accumulated debris near the coop. If feed disappears too fast from the feeder, or if you see holes or trails in the litter, or if you surprise one in the flesh at the feeder at dusk, discuss traps or poison with the people at your feed store. Follow instructions and be careful with poison.

DISEASES

Most diseases your chickens will get are of the respiratory variety. Their symptoms are uncannily like those of the common cold: nasal discharge, sneezing, coughing, and general malaise. Like the spread of a cold, spread of disease is almost always introduced into a healthy flock from an outside source. There is some risk from wild birds coming to feed (and defecate) at your chicken's feeder, but more often the risk comes from chickens you bring in from someone else's flock to join your own.

Some diseases don't cause symptoms at every stage of infection, so if you bring a new bird in, it's hard to tell how healthy it really is. You should quarantine new birds somewhere by themselves for a full month to see if they are going to keel over and die of unknown causes. You can take your test a step further if you want to be doubly safe. Sometimes diseases that don't cause symptoms or much damage in one flock can devastate another. Put one of your own

A WORD ABOUT ANTIBIOTICS

IF YOU USE ANTIBIOTICS TO TREAT YOUR CHICKENS, you must not eat the meat or eggs from treated birds while they are being given the drugs, or for a period of time after they are no longer being treated, called the withdrawal period. The withdrawal period is the time it takes for the drug residue to leave the bird's system. Read the label to find out how long that is.

There is much concern about drug residues in chicken, but a greater risk to humans is the development of antibiotic-resistant bacteria. In large commercial operations, antibiotics are routinely included in chickens' rations not just to fight infection but to enhance their growth. Routine use of antibiotics causes bacteria to develop resistance much more quickly than they would if antibiotic use were more limited. In some cases, the antibiotics used in commercial poultry production are also used to treat humans or are related to antibiotics used to treat humans. The worry then is that we are creating bacteria in our poultry industry that can cause illness in humans that can't be treated with existing antibiotics. Your intermittent use of antibiotics won't make a big difference in the global situation, but it pays to use them carefully, if at all, even in your own backyard.

Above: Roosters Max, right, and Radio Flyer spread their wings and enjoy a nice late fall day at Lianne Thomashow's home. Thomashow's kids named the birds.

birds (one that you can bear to part with if need be) in with the quarantined birds to see what happens to her. If all are well after a month, you can mix them in with your flock.

If you don't have a place to keep another group of birds, or even a way to segregate groups within your coop, you'll have to rely on your own observation, the reputation of the seller, and luck.

If you do suspect disease, you'll probably want to call call the closest extension poultry specialist or get advice from the local feed store, where you can also get the necessary medication. Still, there are a few names and symptoms an informed keeper needs to know about.

CHRONIC RESPIRATORY DISEASE

This infectious, contagious disease is not usually fatal, but it can affect egg production and make your chickens generally unhappy. It is characterized by sneezing, sniffling, and nasal discharge and can be treated with antibiotics.

Marek's Disease

This causes paralysis, usually in young birds. Hatcheries vaccinate for this before they ship day-old chicks to you. There is no treatment for this disease.

Newcastle Disease

This common and dangerous viral respiratory infection causes sneezing, coughing, and often paralysis. There is no treatment for this once a bird has it, although a vaccine does exist. It can also appear as a mild case that is not life-threatening.

Infectious Bronchitis

This is also a respiratory infection causing sneezing and coughing. It can affect the reproductive organs of young birds and lowers egg production in layers. There is no treatment, although a vaccine exists for it.

Infectious Coryza

This results in watery eyes, foul-smelling nasal discharge, and swollen face and wattles. It can be treated with antibiotics.

Fowl Pox

Fowl pox is a viral disease that causes wart-like eruptions on the chicken's skin and in its mouth. It is spread by mosquitoes as well as direct and indirect contact with infected birds. Vaccination is the only prevention, and there is no treatment.

Bumblefoot

Bumblefoot is a staph infection of a wound in the pad of a chicken's foot that causes swelling and an abscess. You prevent it by making sure there are no splinters in your roost and plenty of litter to cushion the feet. Treat it by washing the foot and leg well, lancing and draining the abscess, rinsing with hydrogen per-

oxide, applying bacitracin ointment, and then wrapping
the foot with a bandage.

PARASITES

Parasites, which can be either internal or external, sound
much worse than they are. They can be easily treated with
the correct medication used conservatively.

INTERNAL PARASITES

Protozoa and worms are the most common kinds of internal parasites. Coccidiosis is a common protozoa that takes
up lodging in the intestines of a chicken. This parasite is
discussed in Chapter 4 because it almost exclusively
affects young chicks. Worms are a different kind of
unpleasantness that adult chickens can pick up just by eating in the grass in the yard. Chickens that have too many
tapeworms or other kinds of worms in their gut may eat
less, lay less, and begin to look miserable. Although the
likelihood of a worm problem is small, a vet can diagnose
a problem with a fecal sample and recommend treatment.

EXTERNAL PARASITES

External parasites means mites and lice. These were mentioned in Chapter 5
because your birds are most likely to acquire them on the bodies of adult birds
brought in from someone else's coop. It is also possible for rats and wild birds to
introduce them to your flock. You should occasionally keep an eye out for tiny
crawling things on your bird, particularly in the fluff feathers and around the

vent, and take a closer look if a bird appears listless or is eating or laying less and has a pale comb.

Lice live on your birds for life. They eat the feathers and dead skin

Above: A Silver-Spangled Hamburg rooster cleans himself. By rolling around on the ground, chickens will also "dust" themselves to keep their feathers clean and free of mites.

and scabs of your chickens and lay eggs at the base of the feathers that hatch in 2 to 7 weeks. Lice are more nuisance than danger, but a serious infestation can cause the chicken to itch violently and pull out her own feathers, and may cause her to eat less and therefore lay fewer eggs.

Mites do not settle for feathers and dead skin—they literally suck the blood from the birds and can cause weight loss and a drop in egg production. Some live in the cracks and crevices of the coop and only climb onto the birds at night to latch on.

The scaly leg mite burrows under the skin on the legs and feet of the chicken to feed there, and a buildup of detritus pushes the scales up and makes it painful eventually for the chicken to walk. The general treatment is to smother the mites by covering the legs with petroleum jelly or dipping them (daily or weekly to get rid of a problem and monthly for prevention) in vegetable or linseed oil. Often the oil is mixed with up to one-third kerosene. Since the mites crawl on the roost to get to the chicken, painting the roost with the oil-kerosene mixture can help. The dust bath is the best way the chicken has to control mites and lice herself. Cedar shavings can help keep them away as well, particularly in the nest of a setting hen, who must spend many hours sitting more or less still on a clutch of incubating eggs and cannot so easily scratch or dust.

Finally, there are poisons and powders designed to combat mites and lice. Consult your local feed store or extension poultry specialist.

CANNIBALISM

Picking is central to what makes a chicken a chicken. From the time they are hatched, chicks know how to use their beaks to grab food and ingest it. Perhaps because their own toes and those of their broodermates look something like small worms, "toe-picking" is sometimes a problem at a very early age. The potential for a problem does not go away when the birds are adults, although toes are no longer of interest. The head, vent, and the area around the base of the tail are the first spots that grown chickens will

Opposite: One of Tom Powers' hens gets her legs crossed while eating in the pen outside of the chicken house.

 Above: A Columbian
Wyandotte hen takes
a dust bath.

peck on their coopmates under certain circumstances, and the results can become bloodily disastrous in a very short time.

If you notice feathers or patches of feathers missing from one or more birds, it may be the molt, or it may be that your rooster has been standing on her back in order to do his job. If you suspect picking, however, try a couple of things.

The reason for cannibalism can be related to nutrition, particularly insufficient protein in the ration, and the stress of overcrowding. Invariably, however, cannibalism is ascribed at least in part to boredom. To complain that a chicken is bored strikes me as unnecessarily anthropomorphic and a little weird. And yet, chickens by nature are curious, exceedingly active, and industrious creatures that during the daylight hours are constantly moving, pecking, exploring, scratching, digging, dusting, laying, turning eggs, chasing moths, pulling worms, and chasing the rototiller, with only a short break now and again to look at you sideways and blink. When they are confined to a 10 by 10 box with a dozen cellmates it is perhaps easy to imagine them trying to fill their time with whatever diversion they can find.

Above: Radio Flyer the rooster wanders around outside of the chicken pen at Lianne Thomashow's home. Since she was a little girl, Thomashow had wanted to have farm animals. She got into animal husbandry soon after moving to the country from the city.

To help resolve this, let the birds outside as much as possible. This gives them more to attract their attention and gives birds lower in the pecking order more places to get away from antagonists closer to the top. Next, if you can find a particular offender or ringleader, remove him or her to solitary confinement or the stew pot. Finally, provide toys. Scraps from the kitchen keep them engaged for a little while, but you can take it a step further by hanging a cabbage or something green and leafy from a string suspended from the ceiling of your coop. The birds like to reach, pick, and shred just about anything.

If this doesn't resolve the problem, you have some other options. There are a number of devices that attach to a bird's beak. One such device is made to shield the beak so one bird can't strike another bird with the sharp point. Another amounts to rose-colored glasses that make any spots of blood harder for the chicken to see. A drop of blood shows up in brilliant contrast on a chicken's skin, and only serves to attract further picking. A simple way of countering this phenomenon is to put a red light in the coop.

You can also try one of the many ointments and preparations available in the feed stores and supply catalogues that (reportedly) taste terrible to chickens. If spread on the wounds of afflicted birds, the taste is supposed to discourage further picking.

A drastic measure, but one sure to work, is beak trimming. Personally, I think that beak trimming is mutilation by any standard; if nothing else, birds with trimmed beaks look ridiculous. That said, if you are desperate, you can remove the last quarter-inch of the upper beak with a pair of toenail clippers. If you remove the lower beak, the bird will have nothing to scoop grain up with and will starve. Only about the last third of the upper beak need be removed to prevent cannibalism. It will grow back in about 6 weeks, and you can trim it again if you think it's necessary.

If nothing else works, you'll have to remove the picker from the flock.

CHAPTER TEN
Children and Chickens

"I t's important to realize that you keep chickens for yourself and not for your children—because no one outgrows the pleasure of finding an egg in the nesting box, still warm."

—THE AUTHOR

For the daily gift of the eggs, for a freezer full of unadulterated homegrown roasting chicken, and for the companionship of these independent and yet companionable creatures, you are willing to establish—and probably enjoy—the rhythms of feeding and cleaning and letting out and shutting back in that are your end of your bargain with chickens.

But as the rewards of right living ripple outward, your children are the next in line to benefit. Small children love to join in with work projects led by a parent. My kids put almost as many nails into our present coop as I did. They are particularly skilled with litter, both spreading fresh shavings as needed and shoveling it all out in the spring.

The surprise and delight of finding eggs, of course, is the highlight. One keeper I know has a 5-year-old neighbor

Opposite: A regal White Leghorn hen poses for a portrait of sorts at Tom Powers' home. Powers said he's always liked chickens, and having fresh eggs is one of the three things he cherishes about living in the country—the others are picking ripe blackberries and digging potatoes out of the warm ground.

 Above: Three-year-old Evan Asher
Sandoe feeds a neighbor's chickens in
the family's barn in Whiting,
Vermont..

that has learned to listen for the particular cackle, the "song," he calls it, that the hens sing when they lay an egg. When visiting, he makes trips every 10 minutes to the coop to listen and look for new eggs. A young visitor to my coop once gently carried an egg away with her, and, according to her mother, didn't let it out of her sight for a week. Another friend has encouraged her girls to collect the eggs, wash them, and put them in the cartons to be sold. The girls then get to keep the egg money.

A certain amount of care needs to be taken when introducing children to chickens— and vice versa. Although hardy and self-sufficient, chickens (particularly young chicks) can suffer stress if children handle them too roughly or make too much loud noise or move too quickly or chase them or otherwise treat them as anything other than a fellow creature worthy of respect. The flip side is that children are sensitive, too. One keeper I know once had twice as many layers as comfortably fit in her coop,

Above: Owner Tom Powers decorated his chicken house, with a sculpture to "amuse the chickens," he said. "Of course, they paid no attention."

and the overcrowding caused a general nervousness of the birds that her children picked up on. The coop wasn't a pleasant place to be for the kids or the chickens.

Because children are sensitive, it can be difficult for a parent to feel comfortable involving them with butchering, but children will let you know how much they want to be involved. On a recent butchering day, my 7-year-old was comfortable viewing the killing and processing from a distance of about 30 yards, and his 5-year-old brother had to be reminded often to keep back far enough from the killing cone not to get blood spattered on him. The chicken feet in particular fascinated the younger one, and when he tried to enter the kitchen with one in each hand to show them off, his mother made a point of ushering him back outside again.

A friend with two sons felt strongly enough that they should know where their food comes from that when she heard that we were going to butcher chickens on an upcoming Saturday, she made sure to be there with kids in tow to help out. Her boys were mildly interested for a few minutes before moving on to another activity, but I think I know what the mom was after. Another chicken-raising mother I know explained it to me once this way. She wants her kids to be familiar and easy with the basic elements of the cycle of life in this world: hatching, feathering out, growing up, moving outside, fighting, the establishment of pecking orders, laying eggs, and finally death, whether it happens by our own hands, or by a hungry raccoon, or whether one morning we find one on its back in the litter with its legs sticking up, dead for no apparent reason. She says that these things don't frighten her children—they take them more as a matter of course, as occasions of consequence and gravity and certainly curiosity, but not reasons to be afraid.

In the meantime, whatever the developmental or metaphysical or emotional or political benefits are—impossible to measure or sometimes even describe—chickens are just good to have around. One evening last spring after I had brought the chicks home from the post office and got them watered and fed in their cardboard box in the kitchen, my youngest picked one up carefully, trying not to squeeze too hard. It cheeped once or twice and then settled down in his cupped hands as if he were Saint Francis, and the two of them went to sleep in front of the woodstove.

Above: A chicken boosts herself up into the coop at Tom Powers' home. Powers said that as long as they don't misbehave around the rest of the flock, chickens have permanent tenure at his place.

Above: A Buff Orpington hen strolls by a pair of Bourbon Red turkeys raised by Jim Peavey. The Bourbon Red are a rare breed Peavey is raising for breeding.

APPENDIXES

SELECTED CHICKEN BREEDS BY CLASS
(WHERE THE BIRDS ORIGINATED)

AMERICAN	MEDITERRANEAN	ENGLISH	CONTINENTAL
Wyandotte	Ancona	Orpington	Houdan
Plymouth Rock	Leghorn	Cornish	Faverolle
Rhode Island Red	Andalusian	Australorp	Hamburg
New Hampshire			

SELECTED BREEDS BY SPECIALTY

DUAL PURPOSE	LAYERS	MEAT BIRDS
Dominique	Leghorn	Brahma
Houdan	Buckeye	Cochin
Plymouth Rock	Chantecler	Cornish
Sussex	Delaware	Orpington
Orpington	Dominique	
Wyandotte	Holland	
Australorp	Java	
Rhode Island Red	Wyandotte	
New Hampshire	Langshan	
Langshan	Australorp	
	Dorking	

 Above: One of Carrie Maynard's chickens takes a walk in the woods, to see what she can find.

SELECTED BREEDS BY TEMPERAMENT

CALM	NERVOUS
Cochin	Hamburg
Silkies	Sebright
Cornish	Buttercup
Dorking	Lakenvelder
Orpington	Rhode Island Red
Plymouth Rock	New Hampshire

Selected Breeds by Hardiness

Winter Hardy (More Fully Feathered)	Less So
Brahmas	Andalusian
Orpingtons	Leghorn
Cochins	Minorca
New Hampshire	Hamburg
Reds	Buttercup
Rocks	Naked Neck

HATCHERIES

ALABAMA

Dixie Poultry Farm
PO Box 506
Saraland, AL 36571
866-868-6069
Fax: 1-251-866-0525
http://www.dixiepoultryfarm.com/

CALIFORNIA

Belt Hatchery
7272 S. West Ave.
Fresno, CA 93706
559-264-2090
http://www.belthatchery.com/

CONNECTICUT

Hall Brothers Hatchery
PO Box 1026
Norwich, CT 06360
860-886-2421
Fax: 860-889-6351

Opposite: Carrie Maynard's Silver-Pencilled Wyandotte rooster is considered to be a rare breed. It is one of two Maynard has in her barnyard, and is the top rooster in the pecking order.

GEORGIA

K & L Poultry Farm
772 Morris Road
Aragon, GA 30104
706-291-1977
Fax: 706-294-0110
http://www.klpoultryfarm.com

IOWA

Decorah Hatchery
406 W. Water Street
Decorah, IA 52101
563-382-4103
http://www.decorahhatchery.com

Hoover's Hatchery
PO Box 200
Rudd, IA 50471
800-247-7014
http://www.hoovershatchery.com

Murray McMurray Hatchery
Box 458, 191 Closz Drive
Webster City, IA 50595-0458
800-456-3280
http://www.mcmurrayhatchery.com/

Sandhill Preservation Center— Heirloom Seeds and Breeds
1878 230th Street
Calamus, IA 52729
563-246-2299

Idaho

Dunlap Hatchery
Box 507
Caldwell, ID 83606-0507
208-459-9088

Michigan

Townline Hatchery
PO Box 108
Zeeland, MI 49464
616-772-6514
Fax: 616-772-2969
info@townlinehatchery.com
http://www.townlinehatchery.com/

Minnesota

Stromberg's
Box 400
Pine River, MN 56474-0400
800-720-1134
http://www.strombergschickens.com/

Missouri

C. M. Estes Hatchery, Inc.
PO Box 5776
Springfield, MO 65802
417 862-3593
http://www.esteshatchery.com

Cackle Hatchery
PO Box 529
Lebanon, MO 65536
417-532-4581
http://www.cacklehatchery.com/

Heartland Hatchery
Rt. 1 Box 177-A
Amsterdam, MO 64723
660-267-3679
http://www.heartlandhatchery.com/

Marti Poultry Farm
PO Box 27
Windsor, MO 65360-0027
660-647-3156

McKinney & Govero Poultry
4717 Highway B
Park Hills, MO 63601
573-518-0535 or 573-431-4841
Gdmckinn@i1.net
http://www.mckinneypoultry.com

Mississippi

Yoder's Hatchery
256 Jarrell Rd.
Kokoma, MS 39643
601-736-7800

North Carolina

Seven Oaks Game Farm
7823 Masonboro Sound Road
Wilmington, NC 28409
910-791-5352
http://www.poultrystuff.com

NEBRASKA

Larry's Poultry Equipment & Hatchery
PO Box 629
Scottsbluff, NE 69363
800-676-1096
http://www.larryspoultry.com

NEW MEXICO

Privett Hatchery
PO Box 176
Portales, NM 88130
800-545-3368
http://www.yucca.net/privetthatchery/

OHIO

Allen's Poultry & Gamebird Farm
2165 Alex White Road
Jackson, OH 45640
740-820-4507

Eagle Nest Poultry
Box 504
Oceola, OH 44860
419-562-1993

Mt. Healthy Hatcheries
9839 Winton Road
Mt. Healthy, OH 45231
800-451-5603
Fax: 513-521-6902
http://www.mthealthy.com

OREGON

Shank's Hatchery
17874 Shank Rd NE
Hubbard, OR 97032-9733
800-344-2449
Fax: 503-981-7215
http://home.1stpage.com/1stlook/
pages/standard.cfm?url=shanks

PENNSYLVANIA

The Easy Chicken Poultry and Supply
Scott and Kelly Shilala
R.D. 4 Box 464
DuBois, PA 15801
814-583-5374
http://shilala.homestead.com

Hoffman Hatchery, Inc.
P. O. Box 129
Gratz, PA 17030
717-365-3694
http://www.hoffmanhatchery.com/

Moyer's Chicks, Inc.
266 East Paletown Rd.
Quakertown, PA 18951
215-536-3155
Fax: 215-536-8034
http://www.moyerschicks.com

Reich Poultry Farms, Inc.
1625 River Road
Marietta, PA 17547
717-426-3411

SOUTH DAKOTA

Inman Hatcheries
PO Box 616
Aberdeen, SD 57402-0616
800-843-1962
http://www.inmanhatcheries.com/

TEXAS

Ideal Poultry Breeding Farms, Inc.
PO Box 591P
Cameron, TX 76520-0591
254-697-6677
Fax: 254-697-2393
http://www.ideal-poultry.com/

WASHINGTON

Harder's Hatchery
624 N. Cow Creek Rd.
Ritzville, WA 99169
509-659-1423

Phinney Hatchery, Inc.
1331 Dell Avenue
Walla Walla, WA 99362-1023
509-525-2602

WISCONSIN

Sunnyside Inc. of Beaver Dam
Hatchery Division
PO Box 452
Beaver Dam, WI 53916
920-887-2122

Utgaard's Hatchery
Box 32
Star Prairie, WI 54026
715-248-3200
Fax: 715-248-7410

EQUIPMENT AND SUPPLIES

CALIFORNIA

Morton Jones
P.O. Box 123
Ramona, CA 92065
800-443-5769

CONNECTICUT

Farmtek
1395 John Fitch Blvd
South Windsor, CT 06074
800-327-6835
Fax: 800-457-8887
farmtek@farmtek.com
http://www.farmtek.com/

FLORIDA

Double R Discount Supply
4036 Hield Rd NW
Palm Bay, FL 32907
321-259-9465
http://www.dblrsupply.com

GEORGIA

G.Q.F. Mfg. Co.
PO Box 1552
Savannah, GA 31498
912-236-0651

IOWA

Brower Equipment, Inc.
Highway 16 West, PO Box 2000
Houghton, IA 52631
800-553-1791 or 319-469-4141
Fax: 319-469-4402

Premier1
2031 300th St.
Washington, IA 52353
800-282-6631 Fax: 800-346-7992
http://www.premier1supplies.com/

KANSAS

Smith Poultry & Game Bird Supplies
14000 West 215th Street
Bucyrus, KS 66013
913-879-2587
http://www.poultrysupplies.com/

MICHIGAN

Cutler's Pheasant Supply, Inc.
3805 Washington Road
Carsonville, MI 48419
810-657-9450
http://www.cutlersupply.com

NORTH CAROLINA

Patterson's Poultry & Pet Supplies
P.O. Box 39
Wallburg, NC 27373-0039
336-769-4392

OREGON

Kemp's Koops
3560 W 18th Ave.
Eugene, OR 97402
http://www.poultrysupply.com/

TENNESSEE

Rocky Top General Store
PO Box 1006
Harriman, TN 37748
423-882-8867
Fax: 423-882-9056

TEXAS

Randall Burkey Co., Inc
117 Industrial Drive
Boerne, TX 78006
800-531-1097
http://www.randallburkey.com

VIRGINIA

Egganic Industries
3900 Milton Hwy
Ringgold, VA 24586
800-783-6344

Shenandoah Manufacturing Co., Inc.
PO Box 839
Harrisonburg, VA 22802
800-476-7436
Fax: 540-434-7436
http://www.shenmfg.com/poultry/

WISCONSIN

Nasco Farm & Ranch
901 Janesville Ave.
Fort Atkinson, WI 53538
800-558-9595
Fax: 414-563-8296
http://www.enasco.com

COOPERATIVE EXTENSION OFFICES

For each state there is listed an address and phone where you can contact an extension poultry specialist, and a web link that will take you to a listing of county extension offices in your state.

ALABAMA

Poultry Science Department
Auburn University
Auburn, AL 36849
334-844-2613
http://www.aces.edu/counties/

ALASKA

University of Alaska
P.O. Box 756180
Fairbanks, AK 99775
907-474-7083
http://www.uaf.edu/coop-ext/offices/

ARIZONA

University of Arizona
Tucson, AS 85721
520-621-1980
http://ag.arizona.edu/extension/
counties/

ARKANSAS

Department of Agriculture
University of Arkansas Pine Bluff
P.O. Box 82
Pine Bluff, AR 71601
501-543-8526
http://www.uaex.edu/director/
list4.asp

CALIFORNIA

Department of Avian Sciences
UC Davis, CA 95616
916-752-3513
http://danr.ucop.edu/danrdir/
uccequery.cfm

COLORADO

Department of Animal Sciences
Colorado State University
Fort Collins, CO 80523
970-491-7803
http://www.ext.colostate.edu/

CONNECTICUT

Department of Animal Science,
University of Connecticut
P.O. Box U-40
3636 Horsebarn Road Ext.
Storrs, CT 06268
860-486-1008
http://www.lib.uconn.edu/CANR/
ces/offices.html

DELAWARE

Poultry Research Lab
Route 2, P.O. Box 48
Georgetown, DE 19947
302-856-7303
http://ag.udel.edu/extension/

FLORIDA

Poultry Science Department
P.O. Box 110920
University of Florida
Gainesville, FL 32611
352-392-1931
http://www.ifas.ufl.edu/www/
extension/cesmap.htm

GEORGIA

University of Georgia
4 Towers Building
Athens, GA 30602
706-542-1325
http://www.ces.uga.edu

HAWAII

Department of Animal Sciences
University of Hawaii
18000 East-West Road
Honolulu, HI 96822
808-956-8334
http://www2.ctahr.hawaii.edu/
extout/extout.asp

IDAHO

Southeast Idaho Research
and Extension Center
University of Idaho
16952, S. Tenth Ave.
Caldwell, ID 83605
208-459-6365
http://www.uidaho.edu/ag/
extension/district.html

ILLINOIS

Department of Animal Sciences
University of Illinois
132 Animal Science Lab
1207 W. Gregory Dr.
Urbana, IL 61801
217-244-0195
http://web.aces.uiuc.edu/ve/

INDIANA

Purdue University
Department of Animal Sciences
Lilly Hall
West Lafayette, IN 47907
765-494-8009
http://www.anr.ces.purdue.edu/

IOWA

Department of Animal Sciences
Iowa State University
Ames, IA 50011
515-294-4303
http://www.exnet.iastate.edu/
Counties/state.html

KANSAS

Department of Animal Sciences
Kansas State University
Call Hall
Manhattan, KS, 66506
913-532-6533
http://www.oznet.ksu.edu/root/
units.htm

KENTUCKY

Department of Animal Sciences
University of Kentucky
Lexington, KY 40546
606-257-7529
http://www.ca.uky.edu/county

LOUISIANA

Department of Poultry Science
Louisiana State University
Baton Rouge, LA 70803
504-388-4481
http://www.agctr.lsu.edu/parish/
lcesmap/lcesmap.htm

MAINE

Animal And Veterinary Sciences
University of Maine
127 Hitchner Hall
Orono, ME 04469
207-581-2768
http://www.umext.maine.edu

MASSACHUSETTS

Department of Veterinary and Animal
    Sciences
University of Massachusetts
Amherst, MA 01003
413-545-2312
http://www.umass.edu/umext/
locations.html

MICHIGAN

Animal Science Department
Michigan State University
102 Anthony Hall
East Lansing, MI 48824
517-353-2906
http://www.msue.msu.edu/msue/
ctyentpg/ctyunits.html

MINNESOTA

Department of Avian Sciences
120 Peters Hall
University of Minnesota
St. Paul, MN 55108
612-624-4928
http://www.extension.umn.edu/
offices/

MISSISSIPPI

Poultry Science Department
Mississippi State University
P.O. Box 5188
Mississippi State, MS 39762
601-325-3416
http://msucares.com

MISSOURI

Animal Sciences Department
S105 Animal Science Center
University of Missouri
Columbia, MO 56211
573-882-6658
http://outreach.missouri.edu/regions/

MONTANA

Montana State University
418 Mineral Avenue
Libby, MT 59923
406-293-7781
http://extn.msu.montana.edu/
about_us/counties/counties.html

NEBRASKA

Animal Science Department
University of Nebraska
P.O. Box 830908
Lincoln, NE 68583
402-472-6451
http://ianrwww.unl.edu/ianr/
coopext/recenters.htm

NEVADA

University of Nevada
Reno, NV 89557
702-297-2184
http://www.nce.unr.edu/

NEW HAMPSHIRE

Department of Animal and
   Nutritional Sciences
University of New Hampshire
Kendall Hall,
Durham, NH 03824
603-862-2247
http://ceinfo.unh.edu/Office.htm

NEW JERSEY

Animal Science Department
Cooks College/Rutgers State University
Bartlett Hall
New Brunswick, NJ 08903
732-932-9793
http://www.rce.rutgers.edu

NEW MEXICO

Department of Animal
   and Range Science
New Mexico State University
P.O. Box 30003,
Las Cruces, NM 88003
505-646-3016
http://www.cahe.nmsu.edu/ces/
map.html

NEW YORK

Animal Science Department
Cornell University
Morrison Hall
Ithaca, NY 14853
607-255-8143
http://www.cce.cornell.edu

NORTH CAROLINA

Department of Poultry Science
North Carolina State University
P.O. Box 7608
Raleigh, NC 27695
919-515-5391
http://www.ces.ncsu.edu/counties/

NORTH DAKOTA

Animal and Range Science
North Dakota State University
Hultz Hall
Fargo, ND 58105
701-237-7691
http://www.ag.ndsu.nodak.edu/
ctyweb.htm

OHIO

Animal Sciences Department
Ohio State University
Plumb Hall
Columbus, OH 43210
614-728-6220
http://www.ag.ohio-state.edu/
distcoun.html

OKLAHOMA

Animal Science Building
Oklahoma State University
Room 201
Stillwater, OK 74078
405-744-9293
http://countyext.okstate.edu/

OREGON

Poultry Science Department
Oregon State University
Withcombe Hall
Corvallis, OR 97331
541-737-2254
http://osu.orst.edu/extension/
home/county.html

PENNSYLVANIA

County Extension Office
1383 Arcadia Rd Rm 1
Lancaster, PA 17601
717-394-6851
http://hortweb.cas.psu.edu/MG/
counties.html

RHODE ISLAND

Department of Fisheries, Animals, and
Veterinary Science, Peckham Farm
University of Rhode Island
Kingston, RI
401-874-2072
http://www.uri.edu/ce/

SOUTH CAROLINA

Poultry Science Department
Clemson University
129 P&AS Building
Clemson, SC 29634
864-656-4026
http://virtual.clemson.edu/groups/
extension/clmsites.htm#
CountyExtensionOffices

SOUTH DAKOTA

Animal Science Complex
South Dakota State University
P.O. Box 2175
Brookings, SD 57007
605-693-3484
http://www.abs.sdstate.edu/county/

TENNESSEE

Animal Science Department
PO Box 1071
University of Tennessee
Knoxville, TN 37901
423-974-7351
http://www.utextension.utk.edu

TEXAS

Poultry Science Department 2472
Texas A&M University
College Station, TX 77843
409-845-4318
http://taexhr.tamu.edu/dir/
dircnty.htm

UTAH

Animal and Veterinary Sciences
Utah State University
Logan, UT 84322
801-797-2145
http://extension.usu.edu/coop/

VERMONT

Department of Animal Sciences
Carragan Hall
University of Vermont
Burlington, VT 05405
802-656-2074
http://ctr.uvm.edu/ext/regions.htm

VIRGINIA

Department of Animal &
Poultry Sciences
Virginia Polytechnic Institute
Blacksburg, VA 24061
540-231-5087
http://www.ext.vt.edu/offices/

WASHINGTON

Animal Sciences Department
Puyallup Research And Extension
Center
7612 Pioneer Way East
Puyallup, WA 98371
253-445-4536
http://gardening.wsu.edu/text/
opp.htm

WEST VIRGINIA

Division of Animal and Veterinary
Science
Agricultural Sciences Building
University of West Virginia
Morgantown, WV 26506
304-293-2406
http://www.wvu.edu/~exten/Depart
ments/county.htm

WISCONSIN

Poultry Science Department 260
Animal Sciences, 1675 Observatory Dr
University of Wisconsin
Madison, WI 53706
608-262-9764
http://www.uwex.edu/ces/cty/

WYOMING

Cooperative Extension Service
University of Wyoming
Laramie, WY 82071
307-766-3100
http://www.uwyo.edu/ces/ext3.htm

ONLINE POULTRY RESOURCES

Websites come, go, and move regularly, but I have listed here three kinds of sites that I have found most useful and that appear to be relatively stable. A number of land grant universities with larger poultry science programs work with the extension service to publish articles on poultry-related issues on the Internet, and a number of those sites are listed here. In addition, I have listed a few sites that themselves include large lists of links to other poultry-related web sites, and therefore serve as good starting points for your internet search. Finally, there are names, addresses and websites listed here of a few organizations that may be of interest.

EXTENSION PUBLICATIONS ON THE WEB

Oklahoma State University:
http://www.ansi.okstate.edu/exten/poultry/

Mississippi State University:
http://www.msstate.edu/dept/poultry/msupubs.htm

Virginia Cooperative Extension:
http://www.ext.vt.edu/cgi-bin/WebObjects/Docs.woa/wa/getcat?cat=ir-lpd-pou

http://www.apsc.vt.edu/Faculty/Clauer/clauer.html

University of Georgia:
http://www.ces.uga.edu/ces/pubs.html

Purdue University:
http://ag.ansc.purdue.edu/poultry/extensio.htm

North Carolina State University:
http://www.ces.ncsu.edu/depts/poulsci/

University of Maryland:
http://www.agnr.umd.edu/MCE/Publications/Category.cfm?ID=3&top=32

University of Florida:
http://www.dps.ufl.edu/Poultry/Extension/poulpubs.htm

Auburn University:
http://www.aces.edu/department/extcomm/publications/anr/anrps.php

Other Poultry Links

Feathersite
An on-line zoological garden of domestic poultry, including photos, video and information about various breeds of fowl.

http://www.feathersite.com/

PoultryNet
Georgia Tech sponsors PoultryNet, which calls itself "the largest poultry search engine in the world."

http://poultrynet.gatech.edu/

Texas A&M University
They have one of the better pages of poultry links.

http://posc.tamu.edu/library/dother.html

Missouri Poultry Youth Page

http://www.asrc.agri.missouri.edu/poultry/4-h96.htm

Chicken Feed
Sources of natural chicken feed, knowledge about traditional ways of raising chickens around the world and in old times, putting health before profit in raising and feeding chickens.

http://www.lionsgrip.com/chickens.html

The Coop
The Coop is dedicated to all the folks around the world that raise, breed or show poultry, waterfowl, gamebirds and related species.

http://www.the-coop.org/index.html

ORGANIZATIONS

American Poultry Association
133 Millville Street
Mendon, MA 01756
(508) 473-8769
http://www.ampltya.com/

The American Livestock Breeds Conservancy
Box 477, Pittsboro, NC 27313
http://www.albc-usa.org/

American Pasture Poultry Producers Association
5207 70th Street
Chippewa Falls, WI 54729
http://www.apppa.org/

**ATTRA Appropriate Technonlogy
Transfer For Rural Areas**
PO Box 3657
Fayetteville, Arkansas 72702
http://www.attra.org/

GLOSSARY

BANTAM	A smaller chicken, about a quarter to a fifth the size of a regular-sized chicken.
BLEACHING	The fading of yellow coloring from the beak, shanks, and vent of a yellow-skinned laying hen.
BLOOM	An invisible protective coating on an egg that helps keep bacteria out of it.
BROILER	A young meat bird, also called a fryer.
BREED	A group of chickens that share various characteristics including comb and plumage style.
BROOD	A batch of chicks; to raise chicks; to set on a clutch of eggs until they hatch.
CANDLE	To view the contents of an egg by shining a light through it.
CANNIBALISM	Chickens pecking each other until they do damage.
CLUTCH	A group of eggs accumulated by a hen to set on and incubate.
COCCIDIOSIS	A disease caused by a parasitic protozoa causing diarrhea and death in young chicks.
COCCIDIOSTAT	Drug used to fight coccidiosis.
COCK	A male chicken or rooster a year or more old.
COCKEREL	A male chicken less than a year old.

Opposite: The author emerges from his chicken coop after checking up on a new set of chicks that arrived a week earlier from the hatchery. At left is a playhouse for his two kids.

Above: A flock of meat chickens curl up for the night with Frye the pig. After the birds were slaughtered, the pig began sleeping in a little bed made by his keeper—Thomashow's layers preferred to sleep separately on their roosts.

COMB	The fleshy red spiked top-knot on a chicken's head
CROP	The pouch in a chicken's esophagus, at the base of its neck, that bulges with feed after the bird has eaten.
CULL	To remove a nonproductive or otherwise troublesome bird from the flock; also, the removed chicken itself.
DEBEAK	To remove part of the top beak to prevent cannibalism.
DROPPINGS	Manure.

Embryo	The developing chicken inside a fertile egg.
Fake egg	An object with the shape, size and weight of an egg left to encourage hens to lay in a particular place.
Fryer	A young meat bird, also called a broiler.
Gizzard	An organ in the digestive system of a chicken that grinds food with grit swallowed by the chicken.
Grit	Sand and pebbles eaten by chickens to grind food in its gizzard.
Hen	A female chicken a year or more old.
Hock	The joint in the chicken's leg between the thigh and the shank.
Hybrid	The offspring of a cock and a hen of different breeds.
Keel	The breast bone of the chicken.
Litter	Biodegradable material such as pine shavings used on the floor and in nesting boxes to absorb moisture and keep housing clean.
Mite	A type of external crawling parasite.
Molt	The annual dropping out and regrowing of a chicken's feathers.
Oviduct	The tube through which an egg travels over the course of its formation until it is laid.
Pasting	Manure sticking to the rear of a young chick.
Pecking order	The social ranking of a group of chickens
Perch	A pole a chicken sleeps on at night, also called a roost.

PLUMAGE	The feathering of the chicken as a whole.
PUBIC BONES	The two bones sticking out from either side of the vent.
PULLET	A female chicken less than a year old.
RATION	The total mix of feeds eaten by a chicken.
ROASTER	A bird for cooking whole, larger than a fryer or broiler.
ROOST	A pole a chicken sleeps on at night, also called a perch.
ROOSTER	A male chicken a year or more old. Also called a cock.
SCRATCH	Whole or cracked grain fed to chickens.
SETTING	The incubation of eggs by a hen.
SEXED	Chicks that have been sorted by sex.
SHANK	The lower leg of a chicken.
SPUR	The sharp points on the back of a rooster's shanks.
STANDARD	Short for the *Standard of Perfection*, published by the APA, which describes the "perfect" bird of each breed. Also, the description for any given breed.
STRAIGHT RUN	New chicks that have not been sexed.
TRIO	A cock and two hens (or a cockerel and two pullets) of the same breed and variety
VENT	The opening at the rear of the chicken where the digestive, urinary, and reproductive tracts end.

BIBLIOGRAPHY

Aldrovandi on Chickens: The Ornithology of Ulisse Aldrovandi (1600) Volume II, Book XIV, translated from the Latin with introduction, contents, and note by L.R. Lind. Norman, Oklahoma: University of Oklahoma Press, 1963.

Damerow, Gail, *Storey's Guide to Raising Chickens*. Pownal, Vermont: Storey Press, 1995.

Daniel, Charles and Page Smith, *The Chicken Book*. Athens, Georgiea: The University of Georgia Press, reprint 2000.

Graves, Will, *Raising Poultry Successfully*. Charlotte, Vermont: Williamson Publishing, 1985.

Jull, Morely A., *Successful Poultry Management*. New York: McGraw-Hill Book Company, 1951.

Lee, Andy and Pat Foreman, *Chicken Tractor: The Permaculture Guide to Happy Hens and Healthy Soil*. Columbus, Pennsylvania: Good Earth Publications, 1998.

Lippincott, William Adams and Leslie E. Card, *Poultry Production*. Philadelphia, Pennsylvania: Lea and Febiger, 1946.

Logsdon, Gene, *The Contrary Farmer*. White River Jct., Vermont: Chelsea Green Publishing, 1993.

Luttman, Rick and Gail, *Chickens in Your Backyard: A Beginner's Guide.* Emmaus, Pennsylvania: Rodale Press, 1976.

Macdonald, Betty, *The Egg and* I. New York, New York: HarperCollins Publishers, 1987

Mercia, Leonard, *Raising Poultry The Modern Way.* Pownal, Vermont: Garden Way Publishing, 1990.

Salatin, Joel, *Pastured Poultry Profits$*. Swoope, Virginia: Polyface Inc., 1994.

INDEX